Medicinal Chemistry and Marine Life

(Volume 2)

Antidiabetic Potential of Marine Life

Authored by

Santhanam Ramesh
Karuna College of Pharmacy
Kerala University of Health Sciences
Palakkad, Kerala
India

Ramasamy Santhanam
Fisheries College and Research Institute
Tamil Nadu Veterinary Animal Sciences University
Thoothukudi, India

&

Arumugam Uma
Directorate of Incubation and Vocational training in
Aquaculture (DIVA)
Tamil Nadu, India
Tamil Nadu Dr J. Jayalalithaa Fisheries University
Tamil Nadu, India

Medicinal Chemistry and Marine Life

(Volume 2)

Antidiabetic Potential of Marine Life

Auhtors: Santhanam Ramesh, Ramasamy Santhanam & Arumugam Uma

ISSN (Online): 3082-8325

ISSN (Print): 3082-8333

ISBN (Online): 978-981-5322-85-9

ISBN (Print): 978-981-5322-86-6

ISBN (Paperback): 978-981-5322-87-3

Published by Bentham Science Publishers Pte. Ltd. Singapore. All Rights Reserved.

First published in 2025.

need for a court order if at any point you breach any terms of this License Agreement. In no event will any delay or failure by Bentham Science Publishers in enforcing your compliance with this License Agreement constitute a waiver of any of its rights.

3. You acknowledge that you have read this License Agreement, and agree to be bound by its terms and conditions. To the extent that any other terms and conditions presented on any website of Bentham Science Publishers conflict with, or are inconsistent with, the terms and conditions set out in this License Agreement, you acknowledge that the terms and conditions set out in this License Agreement shall prevail.

Bentham Science Publishers Pte. Ltd.
No. 9 Raffles Place
Office No. 26-01
Singapore 048619
Singapore
Email: subscriptions@benthamscience.net

BENTHAM SCIENCE

CONTENTS

FOREWORD

It is my privilege to write this Foreword for my former teacher and guide, Dr. Ramasamy Santhanam (an author and former Dean of the Thoothukudi Fisheries College and Research Institute, the first fisheries college in Tamil Nadu functioning under this University), for his book titled " Medicinal Chemistry and Marine Life: Volume 2. Antidiabetic Potential of Marine Life".

Diabetes mellitus, a chronic metabolic disorder, is characterized by a rise in blood glucose levels, and it is considered to be a major health hazard. The International Diabetes Federation (IDF) has reported that the number of people suffering from diabetes may increase to 693 million by the year 2045. Because of the increasing number of diabetic patients and the limited number of antidiabetic drugs, the search for new antidiabetic compounds, especially from marine sources, has attracted much interest from the scientific community. The marine environment is considered to be a vast and relatively unexploited source of antidiabetic compounds, which offer great scope for the development of new drugs.

The present title, the first of its kind, deals with marine life possessing potential antidiabetic compounds. I strongly hope that his publication will serve as a valuable resource for the students and teachers of both fisheries and pharmaceutical sciences, besides serving as a potential guide for drug industries in the development of novel, antidiabetic drugs.

I congratulate the team of authors for their timely contribution.

N. Felix
Tamil Nadu Dr. J. Jayalalithaa Fisheries University
Nagapattinam, Tamil Nadu
India

Preface

Diabetes mellitus is a metabolic disorder that is associated with several life-threatening complications, including atherosclerosis, retinopathy, and nephropathy. Diabetes is usually caused by the interaction of genetic and environmental factors and is characterized by a lack of insulin secretion and insulin resistance, which may lead to metabolism disorders of fat, protein, and carbohydrates. In 2013, it was estimated that more than 382 million people had diabetes, and this number is expected to increase to 500 million by 2030. It is expected that this disease will be the 7th leading cause of death. The current therapies available for the treatment of this disorder mainly include oral antidiabetic drugs and insulin injections. However, it is reported that the continuous use of synthetic drugs may cause many side effects. Therefore, there is an urgent need for safe and efficient antidiabetic drugs for the management of this disorder. Marine biota, such as marine microbes, marine plants, and marine animals, have been found to be promising sources with potent antidiabetic activity.

Though a few books are presently available on the therapeutic potential of marine biota, a comprehensive volume dealing with the antidiabetic potential of the different constituents of marine life has not been published so far. The present book, prepared by the scientists of both pharmaceutical and marine biology disciplines, will be the first of its kind to answer this long-felt need. It deals with aspects such as marine-derived molecules and lead compounds with antidiabetic activity, the antidiabetic potential of marine microorganisms, marine macroalgae, marine plants (seagrass and mangrove plants), marine invertebrates, ascidians, and marine fishes, and antidiabetic potential of marine fishery by-products such as fish oils, fish and shellfish wastes, as well as chitosan and its derivatives. It is hoped that the present publication will be of great use as a standard text-cum-reference for teachers, students, and researchers of various disciplines, such as biomedical sciences, pharmaceutical sciences, marine biology, and fisheries science. It will also be a valuable reference for libraries of colleges and universities and as a guide for the pharmaceutical industries involved in the development of new antidiabetic drugs from marine microbes, marine plants, and marine animals.

Santhanam Ramesh
Karuna College of Pharmacy
Kerala University of Health Sciences
Palakkad, Kerala
India

Ramasamy Santhanam
Fisheries College and Research Institute
Tamil Nadu Veterinary Animal Sciences University
Thoothukudi, India

&

Arumugam Uma
Directorate of Incubation and Vocational training in Aquaculture (DIVA)
Tamil Nadu, India
Tamil Nadu Dr J. Jayalalithaa Fisheries University
Tamil Nadu, India

CHAPTER 1

Introduction

Abstract: The formation of diabetes as a metabolic disorder, the causes of the formation of type 1 and type 2 diabetes, risk factors associated with type 2 diabetes mellitus, and the components of marine biota possessing antidiabetic compounds are dealt with in this chapter

Keywords: Antidiabetic compounds, Diabetes mellitus, Marine biota, Risk factors, Type 1 diabetes mellitus, Type 2 diabetes mellitus.

INTRODUCTION

Diabetes Mellitus (DM) has been reported to be one of the top 10 causes of human death globally. 4.2 million deaths were caused by it in 2019, and by 2045, the number is believed to rise to more than 700 million cases [1]. The incidence of diabetes increases most rapidly in low- and middle-income countries due to changes in lifestyle and an aging population. DM is nothing but a metabolic disorder in which the body cells are unable to use glucose effectively. This situation arises due to either low insulin (Type 1 diabetes) or insulin insensitivity (Type 2 diabetes). Among these two types of DM, the incidence of T2DM is becoming more common and accounts for about 90% of all the cases of diabetes [2]. This diabetic condition is characterized by a fasting blood glucose level higher than 126 mg/dL. DM is also known to cause complications like cardiovascular complications, ulcerations, dyslipidemia, endoplasmic reticulum stress, neuropathy, nephropathy, and retinopathy in affected persons. The management practices of diabetes include boosting of insulin sensitivity, reduction of alpha-glucosidase activity, *etc*.

TYPE 1 DIABETES AND TYPE 2 DIABETES: FACTORS, DIFFERENCES, AND CAUSES

Factors Associated with Type 1 and Type 2 DM

Several factors contribute to the formation of both type 1 and type 2 DM and are given below in Table **1**.

Santhanam Ramesh, Ramasamy Santhanam & Arumugam Uma

Table 1. Factors associated with the type 1 and type 2 DM [3].

Factor	Type 1 DM	Type 2 DM
Family history	Less than 20%	About 60%
Genetic locus	Unknown	Chromosome 6
Age at onset	<35 yrs	>40-45 yrs
Type of onset	Abrupt	Gradual
Body weight	Normal	Obesity/Non-obesity
Frequency of occurrence	10-20%	80-90%
Pathogenesis	Autoimmune damage of β-cells	Impaired insulin secretion and insulin resistance
Blood insulin level	Reduced insulin	Normal or increased insulin
Condition of Islet cells	Insulitis and β-cell destruction	No insulitis and late fibrosis of islets
Clinical management	Insulin and diet	Insulin, diet, oral drugs, and exercise

Differences between Type 1 and Type 2 Diabetes

Several differences have been reported between type 1 diabetes and type 2 diabetes in the causes, onset of symptoms, and treatment. Type 1 diabetes is not caused by diet and lifestyle habits; it is an autoimmune condition that develops suddenly and is caused by genetics or other unknown factors. On the other hand, type 2 diabetes is often found to develop over time and is due to the lack of adequate exercise and obesity, which are the biggest risk factors. The causes of type 1 diabetes and type 2 diabetes are given below [4]:

Causes of Diabetes

Type 1 Diabetes

  i. The immune system of the body is largely responsible for inhibiting foreign invaders, like harmful viruses and bacteria.
 ii. Type 1 diabetes is known to be caused by an autoimmune reaction. In type 1 diabetes patients, the immune system is believed to mistake the body's own healthy cells for harmful bacteria and viruses.
iii. The immune system may destroy the insulin-producing beta cells in the pancreas. Under such conditions, the body may not be able to produce insulin. The destruction of the body's own cells by the immune system may be due to genetic and environmental factors, like exposure to viruses.

Type 2 Diabetes

i. People with type 2 diabetes have insulin resistance. Though the bodies of these people still produce insulin, they are unable to use it effectively.

ii. Several lifestyle factors, such as less activity and obesity, are largely responsible for this type of DM.

iii. Genetic and environmental factors may also play an important role in this type of DM. For example, more insulin produced by the pancreas is not effectively used by the body of people with type 2 diabetes. As a result, glucose accumulates in the bloodstream.

RISK FACTORS ASSOCIATED WITH TYPE 2 DIABETES

Several factors have been reported to be responsible for the development of diabetes mellitus (Fig. **1**). Among them, the modifiable risk factors such as physical inactivity, overweight/obesity, poor dietary habits, hypertension, smoking, and certain medications (*e.g.* glucocorticoids) and non-modifiable risk factors like genetics, family history, race/ethnicity, increasing age (>45), and history of gestational diabetes are important ones [5].

MARINE BIOTA AS A SOURCE OF BIOACTIVE COMPOUNDS

The marine ecosystems are considered to be the vast and relatively unexploited sources of bioactive compounds with high chemical diversity. Such metabolites include sulfated polysaccharides, proteins, polyphenols, sterols, fatty acids, tannins, flavonoids, pigments, *etc.* These compounds have been reported to possess remarkable pharmacological activities.

Exclusive metabolites have remarkable pharmacological activities like anticancer, antioxidant, anti-inflammatory, antihyperlipidemic, antidiabetic, antibacterial, antifungal, antiviral, antihypertensive, anticoagulant, immunomodulatory, neuroprotective, *etc.*

MARINE BIOTA AND THEIR ANTIDIABETICS

Among the different components of marine biota, sponges, corals, bacteria, fungi, mollusks, ascidians, brown algae, red algae, and green algae have been reported to possess antidiabetic agents [3].

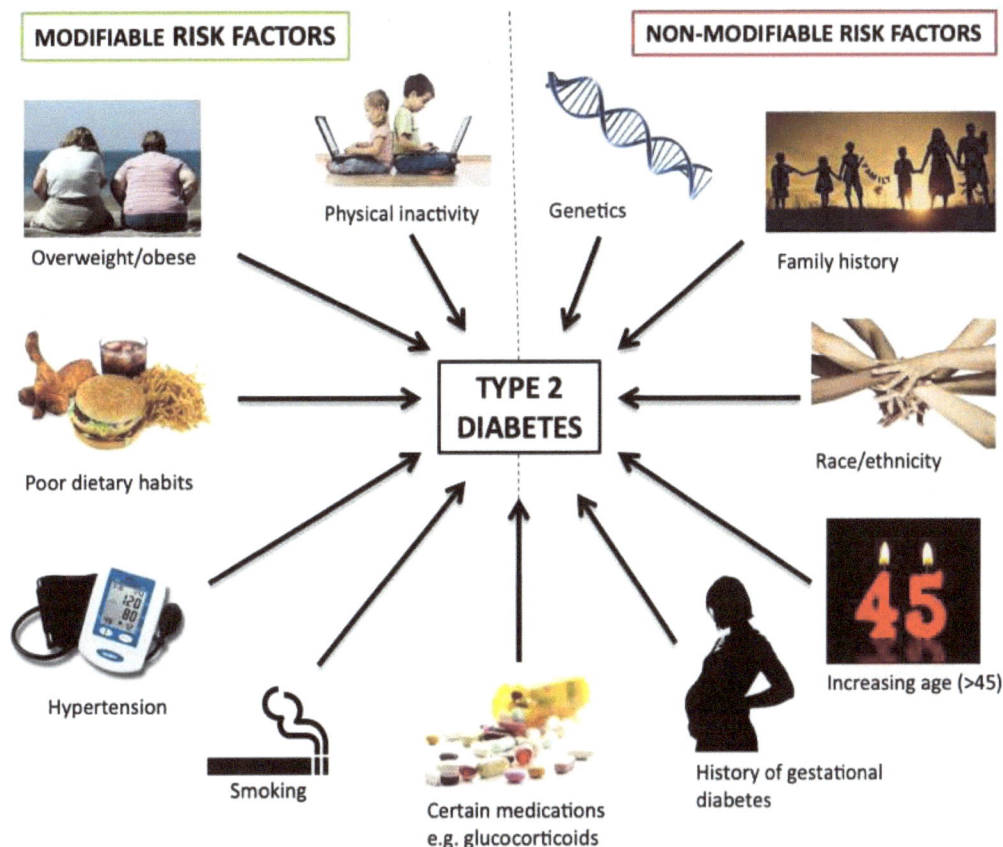

Fig. (1). Risk factors of type 2 diabetes mellitus.
Image credit: Chintha Lankatillake, Tien Huynh and Daniel A. Dias (Applied for permission) Source: Chintha Lankatillake1, Tien Huynh2 and Daniel A. Dias. Understanding glycaemic control and current approaches for screening antidiabetic natural products from evidence-based medicinal plants. Plant Methods (2019) 15:105 https://doi.org/10.1186/s13007-019-0487-8

CONCLUSION

Because of the limited number of anti-diabetic drugs and the increase in the number of diabetic patients, the search for new antidiabetic compounds, especially from marine sources, has attracted much interest from the global scientific community. Although several anticancer, antiviral, and chronic pain-reducing drugs have been derived from marine living resources, sufficient attention has not been paid to the development of antidiabetic drugs from these resources. It is worth mentioning that a terpene compound, *viz.* dysidine, isolated from the sponge Dysidea villosa has entered preclinical trials for the treatment of diabetes (Lauritano and Ianora, 2016). However, there is great scope in the future for the development of anti-diabetic drugs from marine organisms.

CHAPTER 2

Marine Life as a Source of Antidiabetics

Abstract: The properties and chemistry of antidiabetics, the percentage contribution of bioactive compounds by the different components of marine life, and modes of action of antidiabetic compounds derived from certain marine sources are given in this chapter. Further, the role of marine life components in the inhibition of type 2 diabetes, anti-diabetes properties of marine micro and macroorganisms, and marine bioactive compounds such as fucoxanthin, astaxanthin, polyphenol, polysaccharide, krill oil, and fish collagen peptides, as well as their sources and potential applications against diabetes have also been dealt with in this chapter.

Keywords: Antidiabetic compounds, Bioactive compounds, Diabetes, Marine sources, Marine microorganisms, Polysaccharide.

INTRODUCTION

Owing to the limited number of anti-diabetic drugs and worldwide increasing number of diabetic patients, there is an urgent need for the search for new antidiabetic compounds and marine biota that have attracted much interest from the scientific community and offer vast scope in this regard. Among the marine biota, sponges, ascidians, and mollusks have already been reported to yield commercially approved anticancer drugs. For example, the Caribbean sponge *Tethya crypta* has yielded cytarabine (Cytosar-U®, Ara-C, DepoCyt®) to treat non-Hodgkin's lymphoma and acute myelocytic leukemia, vidarabine (Ara-A) derived from the same species is used in the treatment of herpes simplex infections, eribulin (Halaven®) produced by the sponge *Halichondria okadai* is used for the treatment of advanced liposarcoma and metastatic breast cancer, trabectedin (Yondelis®) derived from the tunicate *Ecteinascidia turbinata* has been approved for the treatment of ovarian cancer and tissue sarcomas, and ziconotide (Prialt®), produced by the cone snail *Conus magus* is used in the treatment of severe and chronic pain. It is also worth mentioning that the terpene (Dysidine) derived from the sponge *Dysidea villosa* has already undergone preclinical trials for the treatment of diabetes, and many such antidiabetic lead compounds are under screening. Therefore, the marine biota can offer vast scope in the production of promising antidiabetic compounds in the future [8].

Santhanam Ramesh, Ramasamy Santhanam & Arumugam Uma

Antidiabetics: Properties, Chemistry, and Marine Life Possessing Antidiabetics

The anti-diabetic (anti-glycaemic) properties largely relate to the correction of hyperglycemic and hypoglycemic activities, as well asincreased or reduced secretion of insulin. The factors associated with these phenomena are the inhibition of α-glucosidase, Protein Tyrosine Phosphatase 1B (PTP1B), Dipeptidyl Peptidase IV (DPP-IV), and Glycogen Synthase Kinase-3 beta (GSK-3β), or the protection of beta pancreatic cells. Several bio-actives derived from marine organisms have been reported to possess the aforesaid antidiabetic properties. The chief marine biota possessing antidiabetics and the chemistry of such compounds are given in Tables **1** and **2**.

Table 1. Marine biota possessing antidiabetic properties [3].

Marine Biota	% Contribution of Antidiabetics
Sponges	31
Corals	24
Microbiota	15
Molluscs	6
Ascidians	6
Brown algae	5
Red algae	4
Green algae	1
Others	8

Table 2. Chemical class of antidiabetics derived from marine biota [6].

Marine Biota	Chemistry of Antidiabetics
Marine microorganisms	Antioxidants
Seaweeds	Peptides, amino acids, sterols
Sponges	Peptides
Cnidarians	Phenols
Bryozoans	Alkaloids
Crustaceans	Chitosan, minerals
Mollusks	Polypropionates
Echinoderms	Sterols
Tunicates	Peptides and alkaloids

Modes of Action of Antidiabetic Compounds Derived from Certain Marine Sources

The important antidiabetic compounds derived from major marine resources include phlorotannin, sodium alginate, fucosterol, phenylmethylene hydantoins, n-3 PUFAs, and collagen peptides. The modes of action of these bioactive compounds are given in Table **3**.

Table 3. Antidiabetic effects of marine bioactive compounds of certain marine sources and their modes of action [7].

Marine Source	Antidiabetic Compound	Mode of Action
Ascophyllum nodosum (Alga)	Phlorotannin	Alpha-glucosidase and alpha-amylase inhibitory activities
Laminaria angustata (Alga)	Sodium alginate	Inhibitory action of increase in bold glucose and insulin levels
Pelvetia siliquosa (Alga)	Fucosterol	Decrease in serum glucose level and inhibition of sorbitol in lenses
Ulva rigida (Alga)	Ethanolic Extracts	Reduction in blood glucose levels
Hemimycale arabica (Sponge)	Phenylmethylene hydantoins	Inhibitory action of glycogen synthase kinase-3beta
Fish Oils	n-3 PUFAs	Decrease in blood glucose oxidation; increase in fat oxidation; and maintenance of phosphatidylinositol-3' kinase activity
Wild Fish	Collagen peptides	Reduction in free fatty acids, cytochrome P450, and hs-CRP

Role of Marine Life Components in the Inhibition of Type 2 Diabetes

Among the different components of marine life, microorganisms such as fungi and diatoms (yellow-green algae) and macro-organisms like seaweeds (yellow and red seaweeds), sponges, and ascidians have been reported to possess significant antidiabetic properties as shown in Tables **4** and **5**.

Selected Marine Bioactives and their Potential Applications Against Diabetes

The potential marine bioactives with antidiabetic properties include compounds like fucoxanthin, astaxanthin, polyphenol, polysaccharide, krill oil, and fish collagen peptides [9].

Fucoxanthin

Fucoxanthin (Fig. **1**) is an orange-colored xanthophyll pigment and is derived from the marine macroalgae (brown algae) belonging to the class Phaeophyceae. It is one of the most abundant carotenoid pigments, contributing to about 10% of the total carotenoid production in the marine environment. The popular species of marine algae containing rich fucoxanthin are shown in Table **6**.

Table 4. Marine microorganisms possessing significant antidiabetic properties [8].

Group	Antidiabetic Activity
Bacteria	α-amylase, α-glucosidase, and N-acetyl-glucosaminidase inhibition
Fungi	PTP1B Inhibition
Cyanobacteria (= Blue-green algae)	α-glucosidase inhibition
Microalgae- Green	AGE formation inhibition; α-amylase and α-glucosidase inhibition
Microalgae- Yellow-green	PTP1B Inhibition

Table 5. Marine macroorganisms possessing significant antidiabetic properties [8].

Group	Antidiabetic Activity
Green algae	Reduction of plasma glucose levels in rats
Brown algae	α-amylase, α-glucosidase, and AGE inhibition; Reduction of plasma glucose levels in rats
Red algae	α-amylase, α-glucosidase, sucrase, maltase, and PTP1B inhibition; Aldose reductase and Dipeptidyl peptidase IV inhibition;
Seagrass	Reduction of plasma glucose levels in rats
Marine sponges	α-glucosidase, PTP1B, dipeptidyl peptidase IV, and GSK-3β inhibition
Marine cnidarians (Corals and Sea anemones)	Dipeptidyl peptidase IV inhibition; Reduction of plasma glucose levels in rats
Marine Fishes- Sharks	Promoting insulin secretion; reduction of plasma glucose levels
Byproducts of Marine Fishes- Collagen peptides	Regulation of metabolic nuclear receptors in type-2 diabetes patients
Byproducts of Marine Fishes- Fish oils	Maintaining normal glucose metabolism

Table 6. Marine algae containing rich fucoxanthin [10].

Class/Species	Fucoxanthin Content (mg/g)
Macroalgae; Brown algae (Phaeophyceae)	----
Myagropsis myagroides	9.01
Dictyota coriacea	6.42

(Table 6) cont.....

Class/Species	Fucoxanthin Content (mg/g)
Cystoseira hakodatensis	1.53
Sphaerotrichia divaricata	1.48
Sargassum siliquosum	1.41
Fucus vesiculosus	1.24
Sargassum horneri	1.10
Undaria pinnatifida	1.09
Sargassum duplicatum	1.01
Microalgae; Synurophyceae; *Mallomonas* sp.	26.6
Bacillariophyceae; *Phaeodactylum tricornutum*	24.2
Coscinodiscophyceae; *Odontella aurita*	21.7
Prymnesiophyceae; *Isochrysis* aff. *galbana*	18.2

Antidiabetic Properties of Fucoxanthin

Fucoxanthin is known to possess several physiological functions and biological properties, such as antidiabetic, anti-obesity, antioxidant, antitumor, anti-inflammatory, and hepatoprotective activities, as well as cerebrovascular and cardiovascular protective effects. Among the various bioactivities of this pigment, its antidiabetic properties assume greater importance. Excessive energy intake and accumulation of lipids are known to cause obesity, which elevates insulin resistance in patients with type 2 diabetes. Fucoxanthin has been reported to play a significant role in reducing insulin resistance and blood glucose. Supplementation of 0.2% fucoxanthin was found to reduce plasma insulin levels [9 - 11].

Astaxanthin

Astaxanthin (3,3′-dihydroxy-β, β′-carotene-4,4′-dione) (Fig. **2**) is a red pigment belonging to xanthophyll carotenoids. This natural lipid-soluble compound is widely found in marine microorganisms and the shell portions of marine seafood, such as shrimp, lobster, and crab. The popular marine organisms containing astaxanthin are shown in Table **7**.

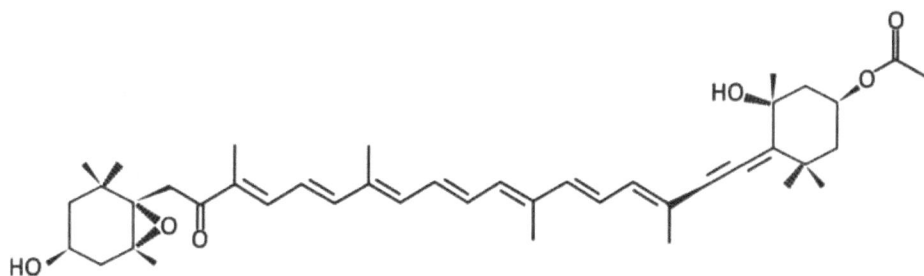

Fig. (1). Fucoxanthin.

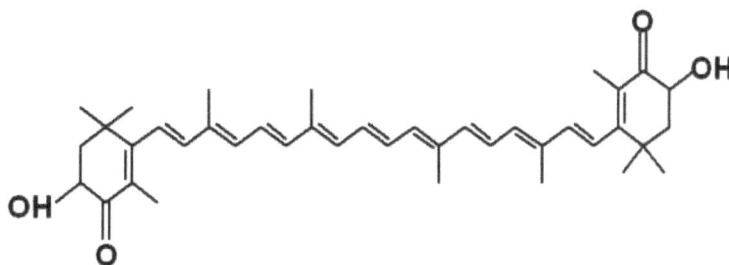

Fig. (2). Astaxanthin.

Table 7. Popular marine organisms containing astaxanthin [12].

Group	Major Species	Astaxanthin (% dw)
Alphaproteobacterium	*Agrobacterium aurantiacum*	0.01
Green algae	*Enteromorpha intestinalis*	0.02
--do--	*Ulva lactuca*	0.01
Red alga	*Catenella repens*	0.02
Caridean shrimp	*Pandalus borealis*	0.12
--do--	*Pandalus clarkia*	0.015

Antidiabetic Properties of Astaxanthin

As a nutritional supplement with antioxidant and anticancer properties, astaxanthin has been reported to prevent diabetes, neurodegenerative disorders, and cardiovascular diseases besides stimulating immunization. The antidiabetic properties of astaxanthin are well known. The oxidative stress levels are invariably very high in patients with diabetes mellitus patients. Further, this stress is often induced by hyperglycemia due to the malfunctioning of pancreatic β-cells. Astaxanthin has been reported to reduce this oxidative stress and improve glucose and serum insulin levels. This pigment has also been found to be a good immunological agent in the recovery of lymphocyte dysfunctions in diabetic rats.

Improved insulin sensitivity has also been reported in experimental rats fed with astaxanthin. Astaxanthin has also been reported to prevent diabetic nephropathy by reducing oxidative stress and renal cell damage [9 - 12].

Phenolics

Among marine metabolites with biological properties, phenols have attracted much attention among researchers due to their potential health benefits in numerous human diseases. Several phenolic compounds derived from marine life have been reported to display bioactivities, which include antidiabetic, antimicrobial, antiviral, anticancer, antioxidant, and anti-inflammatory activities. While polyphenols, simple phenolic acids, and flavonoids are known to help keep one healthy and protect against various diseases, the bromophenolic compounds and phlorotannins are exclusively found in marine sources [9].

Antidiabetic Properties of Phenolic Compounds

The multi-targeted protective effect of marine phenolics on type 2 diabetes mellitus has attracted great interest nowadays. Among the phenolics, the phlorotannins of edible seaweeds have been reported to possess several antidiabetic mechanisms, *viz.* inhibition of starch-digesting enzymes α-glucosidase and α-amylase, inhibition of Protein Tyrosine Phosphatase 1B (PTP1B) enzyme, reduction in glucose levels, lipid peroxidation, modulation of glucose-induced oxidative stress, *etc.*[9]. The antidiabetic activity of major phenolic compounds derived from marine life is shown in Table **8**.

Table 8. Marine phenolics with antidiabetic activity [10].

Source	Marine Group	Major Compounds	Antidiabetic Activity
Ecklonia. cava	Brown alga	Phlorotannins	Inhibition of α-glucosidase
Lessonia trabeculata	Brown alga	Polyphenol-rich extracts	Inhibition of α-glucosidase and lipase activities
Fucus vesiculosus	Brown alga	Phlorotannins	Inhibition of α-amylase and pancreatic lipase
Fucus spp.	Brown algae	Phlorotannins	Inhibition of α-glucosidase and xanthine oxidase
Rhodomela confervoides	Red alga	Bromophenols	Inhibition of PTP1B activity
Lessoniaceae		Phlorotannins	Inhibition of AGE formation
Padina pavonica and *Turbinaria ornata*	Brown algae	Phlorotannins	Inhibition of AGE formation
Ecklonia stolonifera	Brown alga	Phlorotannins	Inhibition of AGE formation

(Table 8) cont.....

Source	Marine Group	Major Compounds	Antidiabetic Activity
Fucus vesiculosus	Brown alga	Polyphenol	No change in postprandial blood glucose and insulin levels
Lessonia trabeculata	Brown alga	Polyphenol	Inhibition of α-glucosidase and lipase
Enteromorpha prolifera	Green alga	Flavonoids	Hypoglycemic effect

Polysaccharides

Polysaccharides are carbohydrate macromolecules and are made of several monosaccharides, which are linked to each other by glycosidic bonds. These compounds have been reported to possess several bioactivities, including hypoglycemic, anticancer, anti-inflammatory, immunomodulatory, and neuro-modulatory. Among the different constituents of marine life, blue-green algae, brown algae, red algae, and sea cucumbers have been reported to possess polysaccharides with hypoglycemic effects [13], as shown in Table **9**.

Table 9. Polysaccharides containing marine life with hypoglycemic effects [13].

Group	Species
Blue-green alga	*Spirulina platensis*
Green alga	*Ulva lactuca*
Brown algae	*Ascophyllum nodosum, Dictyopteris divaricata, Ecklonia maxima, Fucus vesiculosus,Laminaria japonica, Macrocystis pyrifera, Sargassum confusum, Sargassum fusiforme, Sargassum pallidum, Sargassum thunbergia* and *Undaria pinnatifida*
Red algae	*Gracilaria lemaneiformis, Gracilaria opuntia, Kappaphycus alvarezii, Porphyridium cruentum,* and *Porphyra* spp.
Sea cucumbers	*Acaudina molpadioides, Apostichopus japonicus,Cucumaria frondosa, Holothuria leucospilota, Holothuria tubulosa, Isostichopus badionotus, Pearsonothuria graeffei, Stichopus japonicas* and *Thelenota ananas*

Krill Oil

The krill oil derived from the Antarctic krill, *Euphausia superba*, a crustacean zooplankton, has been found to be rich in long-chain omega-3 PUFAs, and several potent antioxidants, such as vitamins A and E, astaxanthin, flavonoid and aglycone. The beneficial bioactivities of this krill oil include antidiabetic, anti-inflammatory, antioxidant, antihyperlipidemic, and antihyperglycemic [16].

Antidiabetic Activity of Krill Oil

Supplementation of krill oil has yielded antidiabetic activity by significantly reducing fasting blood glucose in experimental mice with diet-induced obesity and insulin resistance [9]. Insulin sensitivity was found to be 14% lower with a blend of krill and salmon oil supplementation than with the control oil (Matsuda index: 4.57 compared with 5.33). [14] The supplementation of krill oil for 7-14 days can significantly reduce blood glucose levels in hyperglycemic rats [15]. For instance, while the blood glucose level of placebo group rats was 222.16 mg/dL, the krill oil group registered only 99 mg/dL. Treatment with krill oil gives an improvement in diabetic neuropathy [16]

Marine Collagen Peptides

Marine collagen peptides are low-molecular-weight peptide compounds derived from the skin of certain deep-sea fishes, including chum salmon (*Oncorhynchus keta*), by enzymatic hydrolysis. These peptides are known to possess several multifunctional properties, including anti-diabetic, anti-ulcer, antioxidative, anti-skin aging, and anti-hypertension [17].

Antidiabetic Properties of Marine Collagen Peptides

Marine collagen peptides derived from deep-sea fish have been reported to show protective effects against diabetes by affecting the levels of molecules involved in diabetic pathogenesis. Further, these peptides significantly reduce the level of fasting blood glucose and increase the levels of the insulin sensitivity index and insulin secretion index [9]. The collagen peptides derived from the bone of the fish *Harpadon nehereus* (HNCP) reduced blood glucose and improved glycolipid metabolism in streptozocin-induced type 1 diabetic mice. After 240 mg/kg HNCP treatment, the blood glucose level was found to decrease by 32.8%, while the level of serum insulin increased by 142.0% [17].

CONCLUSION

Marine bioactive compounds are gaining greater attention in novel drugs for the treatment of diseases like cancer, diabetes, and infectious diseases. While the plant and synthetic sources are presently useful in the treatment of diabetes, an ideal drug to combat diabetes and its associated complications is still lacking, and this is largely believed to be due to the diverse nature of diabetes. Marine life is becoming the best choice to meet these challenges as it is still an underutilized, valuable source rich in promising alternative antidiabetic compounds. More detailed investigations are, however, required to provide a valid platform for improving the existing antidiabetic therapeutic strategies.

CHAPTER 3

Antidiabetic Potential of Marine Microbiota

Abstract: Among the marine biota, the macro-organisms have yielded a considerable number of bioactive compounds in the last 50 years of bioprospecting research. Owing to the repeated derivation of known bioactive compounds and reduced number of novel compounds from these macroorganisms, scientists are now trying to concentrate on the less investigated drug sources like marine bacteria and fungi among the microbes, as well as microalgae other than cyanobacteria (blue-green algae) such as yellow-green algae (diatoms) and dinoflagellates, which are known to possess promising bioactive compounds including novel antidiabetics. Further, unlike macroorganisms, microorganisms are known to possess the advantage of sustainable production of considerable quantities of bioactive compounds by large-scale cultivation.

Keywords: Antidiabetic potential, Antarctic lichen, Bacteria, Blue-green algae (cyanobacteria), Fungi, Green algae, Haptophyte alga, Marine microbiota, Ochrophyte algae, Red algae, Yellow-green algae (diatoms).

INTRODUCTION

The marine microbiota constitutes microalgae, including prokaryotic cyanobacteria, eukaryotic microalgae, and endophytic bacterial and fungal diversity. Among these components, certain species of marine microalgae have been reported to serve as functional foods. The bioactive compounds of these microalgae, such as amino acids, peptides, Polyunsaturated Fatty Acids (PUFAs), pigments, scytonemins, pterins, and phenolic compounds, have demonstrated a wide range of therapeutic effects, including anti-inflammatory, antioxidant, and immune-modulatory properties, which help improve insulin sensitivity, thereby potentially alleviating diabetes mellitus. However, the FDA has not approved any microalgae-based antidiabetic products. More research and clinical studies are therefore needed to confirm the antidiabetic potential of these marine microalgae. Apart from these marine microalgae, marine microbes such as the endophytic bacterial and fungal diversity have been extensively studied for their antioxidant, antibacterial, cytotoxic, antidiabetic, and anti-immunosuppressive properties [6 - 8].

Santhanam Ramesh, Ramasamy Santhanam & Arumugam Uma

Antidiabetic Potential of Marine Microalgae

Marine microalgae such as cyanobacteria, yellow-green algae, haptophyte alga, ochrophyte algae, green algae, and red algae have been reported to possess antidiabetic properties.

Antidiabetic Blue-Green Algae (Cyanobacteria)

Aphanothece sp. (Fig. 1)

Global distribution: This picocyanobacterial species has a global distribution, which includes most of the world's oceans.

Ecology: It is known to occur in tropical to subtropical swamps and in brackish-water environments.

Antidiabetics and their mechanisms of action: The extracellular polysaccharide of this species has been reported to possess antidiabetic activity by inhibiting alpha-glucosidase at 8.1% [20].

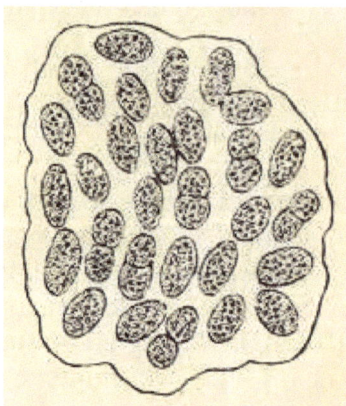

Fig. (1). *Aphanothece* sp Engler, Adolf *et al.* Creative Commons Attribution-Share Alike 4.0 International license.
https://commons.wikimedia.org/wiki/File:Die_Nat%C3%BCrlichen_Pflanzenfamilien_nebst_ihren_Gattunge n_und_wichtigeren_Arten_in%E2%80%A6Fig_49M.jpg.

Arthrospira platensis (= Spirulina platensis)

Global distribution: It is most widely distributed in Asia and Africa.

Ecology: This planktonic filamentous cyanobacterium grows in alkaline, brackish, and saline waters.

Antidiabetics and their mechanisms of action: The crude methanol extracts of this species have shown α-amylase and α-glucosidase inhibitory activities with IC50 values of 13.3 and 9.6 mg/mL, respectively, and percentage values of 96.5 and 97.4, respectively [18]. The oral administration of 80% ethanol extracts of this species at 2.5mg/kg bw to alloxan-induced diabetic rats for 30 days showed a highly hypoglycemic effect [19] (Fig. **2**).

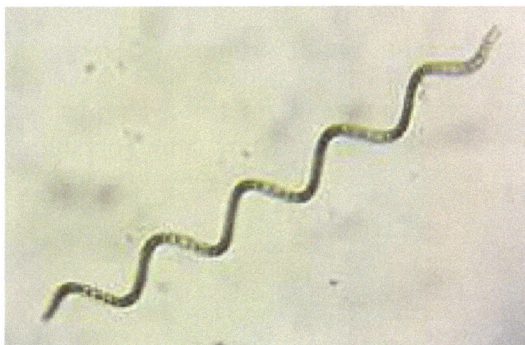

Fig. (2). *Arthrospira platensis*: FarmerOnMars ; Creative Commons Attribution-Share Alike 3.0 Unported license.; https://commons.wikimedia.org/wiki/File:SingleSpirulinaInMicroscope4WEB.jpg.

Chroococcus **sp.** (Fig. **3**)

Global distribution: It is widely distributed across North America.

Ecology: It is commonly seen in eutrophic lakes and ponds. It may also be seen in estuarine environments.

Antidiabetics and their mechanisms of action: The extract of this species has shown alpha-glucosidase inhibition with a percentage value of 13 [20].

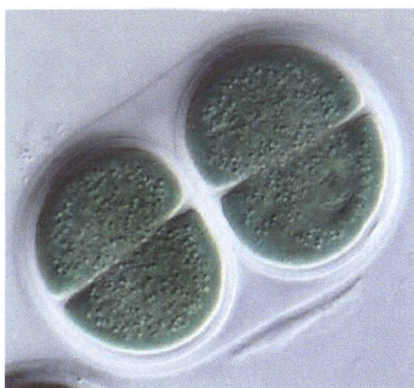

Fig. (3). *Chroococcus* sp.(M. Lorenz, Creative Commons Attribution 4.0 International license. https://commons.wikimedia.org/wiki/File:SAG_36.85_Chroococcus_turgidus_SWES6wk_ML_160818_010_ovl_prime.jpg.

Coelastrella sp. (Fig. 4)

Global distribution: It is widely distributed all around the world, from the tropical to the polar waters.

Ecology: Commonly seen in the rocks of freshwater environments. It may, however, also be seen in estuarine environments.

Antidiabetics and their mechanisms of action: The extract of this species has shown alpha-glucosidase inhibition with a percentage value of 11 [20].

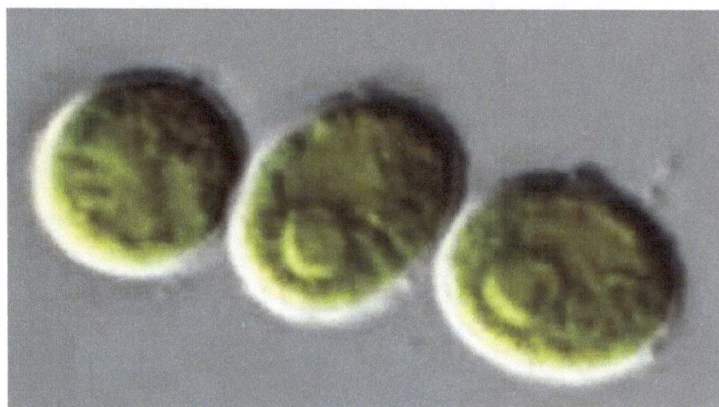

Fig. (4). *Coelastrella* sp Image credit: K. Nayana, M. P. Sudhakar and K. Arunkumar (Reproduced with kind permission).

Source: *Coelastrella* sp. Biorefnery potential of Coelastrella biomass for fuel and bioproducts—a review K. Nayana1 · M. P. Sudhakar2 · K. Arunkumar. Biomass Conversion and Biorefnery https://doi.org/10.1007/s13399-022-02519-9 (Fig.2D); K. Arunkumar rsrnarun@gmail.com; arunnir@yahoo.co.in (to be)

Fischerella sp. (Fig. 5)

Global distribution: It has a cosmopolitan distribution.

Ecology: It is mostly found on moist rocks of freshwater and estuarine environments.

Antidiabetics and their mechanisms of action: The aqueous crude extract of this species shows alpha-glucosidase inhibitory activity at 7.5% [18].

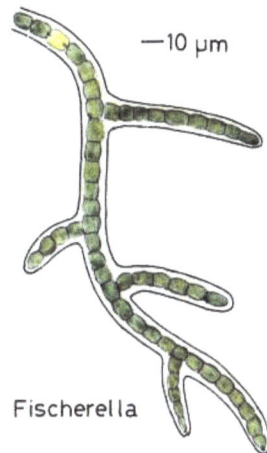

—10 μm

Fischerella

Fig. (5). *Fischerella* sp.-(Pentecost, Allan Creative Commons Attribution-Share Alike 3.0 Unported license. https://commons.wikimedia.org/wiki/File:Cyanobacteriabranchedforms026_Fischerella.jpg.

Leptolyngbya sp. (Fig. 6)

Global distribution: It has a cosmopolitan distribution.

Ecology: It is seen in various ecological habitats, including marine, freshwater, and swamps.

Antidiabetics and their mechanisms of action: The extract of this species has shown alpha-glucosidase inhibition with a percentage value of 2.1 [20].

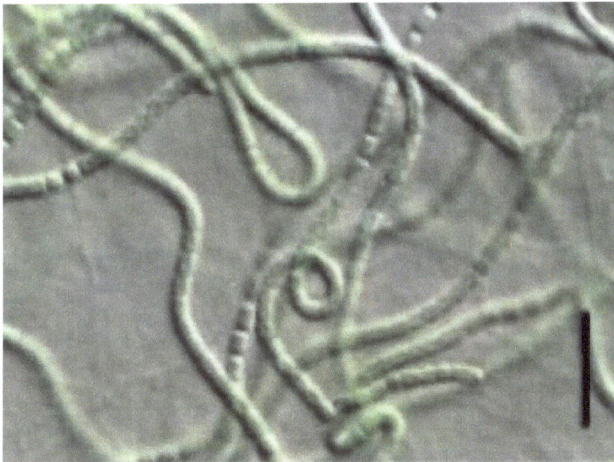

Fig. (6). *Leptolyngbya* sp.(.: Rezeda Z. Allaguvatova, *et al.*, Creative Commons Attribution 4.0 International license. https://commons.wikimedia.org/wiki/File:Leptolyngbya_cf._foveolarum.png

Lyngbya sp. (Fig. 7)

Global distribution: It inhabits coastal and estuarine environments worldwide.

Ecology: It is seen in shallow waters with frequent exposure to diverse environmental stress factors.

Antidiabetics and their mechanisms of action: The extract of this species has shown alpha-glucosidase inhibition with a percentage value of 5.5 [20].

Oscillatoria sp. (Fig. 8)

Global distribution: It is widely distributed over the North American continent.

Ecology: It occurs both in freshwater and marine habitats.

Antidiabetics and their mechanisms of action: The extract of this species has shown alpha-glucosidase inhibition with a percentage value of 9.3 [20].

Fig. (7). *Lyngbya* sp.: NASA, public domain ; https://commons.wikimedia.org/wiki/File:Lyngbya.jpg.

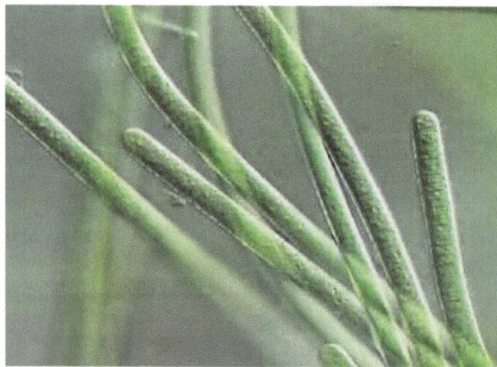

Fig. (8). *Oscillatoria* sp.(.: ja:User:NEON / User:NEON_ja ; Creative Commons Attribution-Share Alike 2.5 Generic license. https://commons.wikimedia.org/wiki/File:Oscillatoria_sp.jpg.

Pseudnabaena sp.

Global distribution: It is widely distributed throughout the temperate zone, where it is known to occur mainly in colder seasons.

Ecology: This species lives in freshwater, brackish water, and marine habitats.

Antidiabetics and their mechanisms of action: The extract of this species has shown alpha-glucosidase inhibition with a percentage value of 14 [20].

Spirulina fusiformis

Global distribution: It is distributed in tropical African countries.

Ecology: It lives in a variety of habitats with diverse environmental conditions.

Antidiabetics and their mechanisms of action: The pigment-protein complex, *viz.* phycocyanin (Fig. **9**), derived from this species shows an anti-hyperglycaemic effect besides reducing the elevated blood glucose level [6].

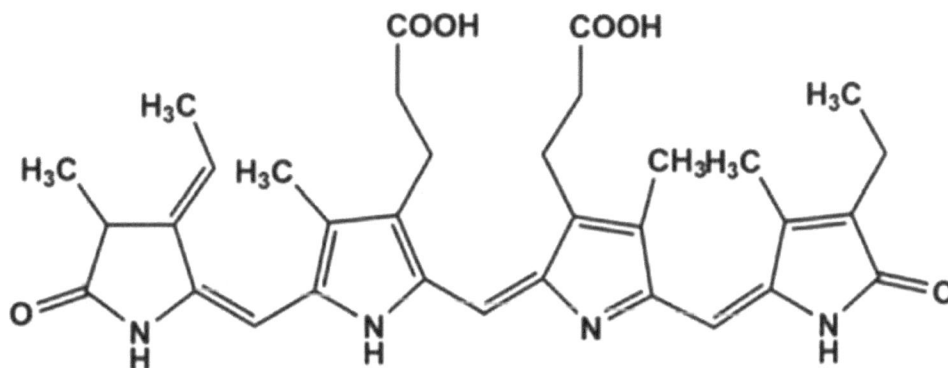

Fig. (9). Phycocyanin.

Spirulina sp.

Antidiabetics and their mechanisms of action: Its butanol crude extract has been reported to exhibit α-glucosidase inhibition with an IC50 value of 23 μg/mL [18].

Synechococcus sp. (Fig. **10**)

Global distribution: It is found distributed in all regions of the oceanic euphotic zone and is ubiquitously seen even in polar waters and deep seas.

Ecology: It is known to prefer the upper, well-lit portion of the euphotic zone.

Antidiabetics and their mechanisms of action: The extract of this species has shown alpha-glucosidase inhibition with a percentage value of 3.1 [20].

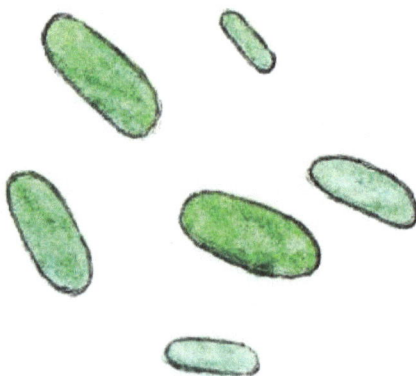

Fig. (10). *Synechococcus* sp: Pentecost, Allan; Creative Commons Attribution-Share Alike 3.0 Unported license.;
https://commons.wikimedia.org/wiki/File:Cyanobacteriaunicellularandcolonial020_Synechococcus.jpg.

Antidiabetic Yellow-Green Algae (Diatoms)

Attheya longicornis

Global distribution: Northern Atlantic regions.

Ecology: It is seen in pelagic environments.

Antidiabetics and their mechanisms of action: The crude extracts of this species have shown antidiabetic activity by inhibiting the protein tyrosine phosphatase 1B (PTP1B) enzyme [8] (Fig. **11**).

Chaetoceros furcellatus

Global distribution: It is an Arctic neritic diatom commonly seen in the Barents Sea.

Ecology: It is seen in a variety of marine habitats.

Antidiabetics and their mechanisms of action: The crude extracts of this species have shown antidiabetic activity by inhibiting the protein tyrosine phosphatase 1B (PTP1B) enzyme [8] (Fig. **12**).

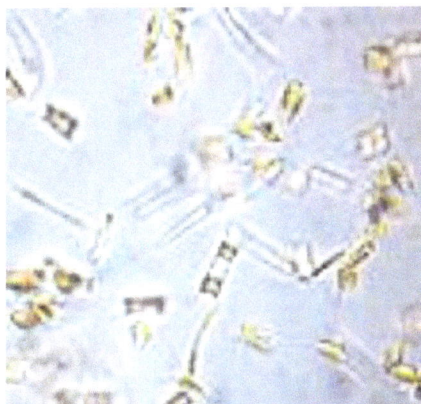

Fig. (11). *Attheya longicornis*. Richard A. Ingebrigtsen, Creative Commons Attribution-Share Alike Attribution-Share Alike 4.0 International , 3.0 Unported , 2.5 Generic , 2.0 Generic and 1.0 Generic license.

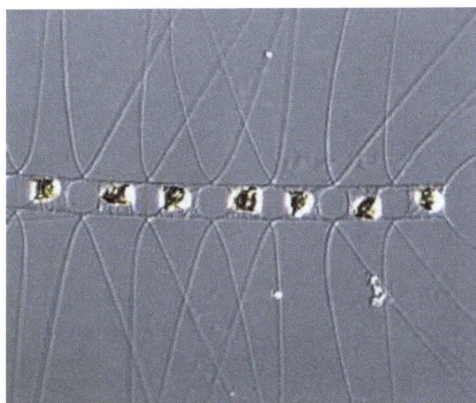

Fig. (12). *Chaetoceros furcellatus*. Yang Li, Creative Commons Attribution 4.0 International license.; https://commons.wikimedia.org/wiki/Tilc:Journal.pone.0168887.g004.tif.

Chaetoceros karianus

Global distribution: Europe, North America, and South America.

Ecology: It is a coastal species.

Antidiabetics and their mechanisms of action: The isomeric oxo-fatty acids *viz.* (7E)-9-oxohexadec-7-enoic acid (1) and (10E)-9-oxohexadec-10-enoic acid (Figs. **13** and **14**) derived from this species have shown antidiabetic activity by inducing antidiabetic gene programs in adipocytes through the upregulation of insulin-sensitizing adipokines [21].

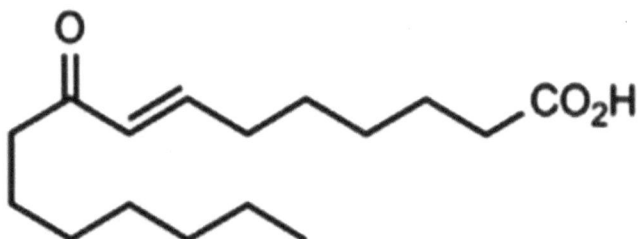

Fig. (13). (7E)-9-oxohexadec-7-enoic acid.

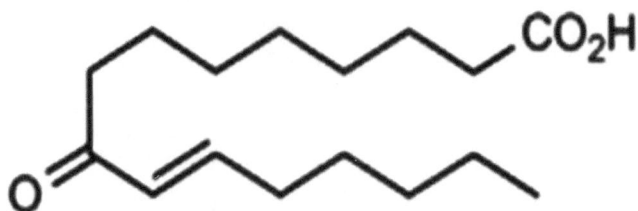

Fig. (14). (10E)-9-oxohexadec-10-enoic acid.

Chaetoceros socialis

Global distribution: It has a cosmopolitan distribution.

Ecology: It inhabits shallow, near-shore areas.

Antidiabetics and their mechanisms of action: The crude extracts of this species cultivated at high temperature–low light have shown antidiabetic activity by inhibiting the protein tyrosine phosphatase 1B (PTP1B) enzyme [8].

Nitzschia laevis

Global distribution: Europe, N. America, S. America, Africa, Southwest Asia, Australia, and New Zealand.

Ecology: It is a shallow, coastal species (Fig. **15**).

Antidiabetics and their mechanisms of action: The crude extracts of this species have been reported to exhibit significant inhibitory effects against the formation of total advanced glycation end products (AGEs), which are believed to be implicated as biomarkers in the development of diabetes. Further, the linoleic, arachidonic, and eicosapentaenoic (EPA) (Figs. **16-18**) fatty acids have been reported to possess strong anti-glycative capacities [8] .

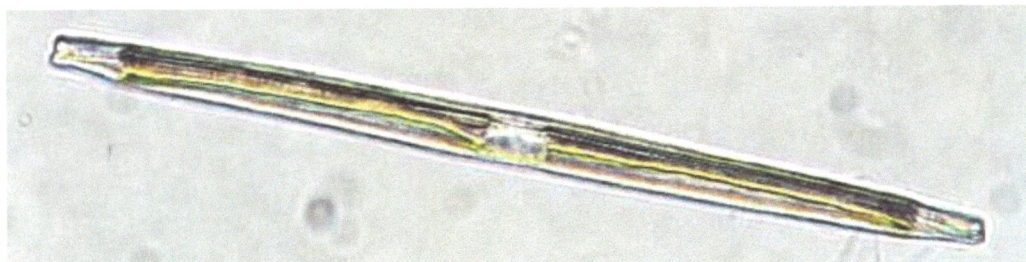

Fig. (15). *Nitzschia* sp.; Kristian Peters; Creative Commons Attribution-Share Alike 3.0 Unported license.; https://commons.wikimedia.org/wiki/File:Nitzschia_sp.jpeg.

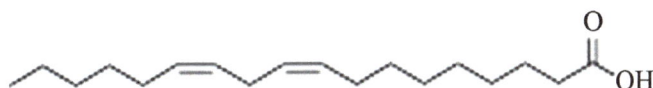

Fig. (16). Linoleic acid.

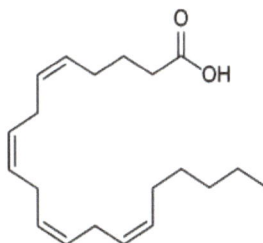

Fig. (17). Arachidonic acid.

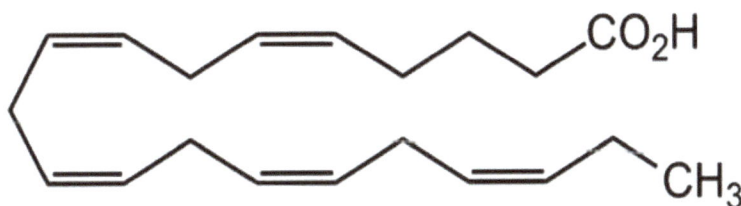

Fig. (18). Eicosapentaenoic acid.

Odontella aurita

Global distribution: North temperate regions.

Ecology: It is seen in a variety of habitats, such as open oceans, estuaries, lagoons, and bays.

Antidiabetics and their mechanisms of action: It is known to contain 7.1 mg g−1 dry cell weight (DCW) of fucoxanthin, which has been reported to be a powerful antidiabetic pigment [22] (Fig. **19**).

Fig. (19). *Odontella aurita.* An, S.M. *et al.*, Description and Characterization of the Odontella aurita OAOSH22, a Marine Diatom Rich in Eicosapentaenoic Acid and Fucoxanthin, Isolated from Osan Harbor, Korea. Mar. Drugs 2023, 21, 563. https://doi.org/10.3390/md21110563 (CC)

Paralia sulcata

Global distribution: It has a cosmopolitan distribution.

Ecology: It occurs in neritic and littoral zones of the marine environment, and it is also occasionally seen in the plankton.

Antidiabetics and their mechanisms of action: Its water and butanol extracts have been reported to stimulate the secretion of glucose-dependent insulinotropic peptide (GIP) *viz.* glucagon-like peptide 1 (GLP-1), which helps the release of insulin by pancreatic islet cells besides modulating β-cell proliferation [23] (Fig. **20**).

Fig. (20). *Paralia sulcata.* Jmpost ; Creative Commons Attribution-Share Alike 3.0 Unported license.; https://commons.wikimedia.org/wiki/File:Paralia_sulcata_diatom.tif.

Phaeodactylum tricornutum

Global distribution: It is commonly found distributed in Europe and Australia.

Ecology: This pennate diatom has both pelagic and benthic marine habitats.

Antidiabetics and their mechanisms of action: Its fucoxanthin has been reported to exhibit PTP1B inhibitory activity with an IC50 value of 4.8 μM [24]. Its fucoxanthin extract displayed both α-amylase and α-glucosidase inhibitory activities with IC50 values of 0.68 and 4.75 mmol/L, respectively [18] (Fig. **21**).

https://en.wikipedia.org/wiki/Attheya_longicornis

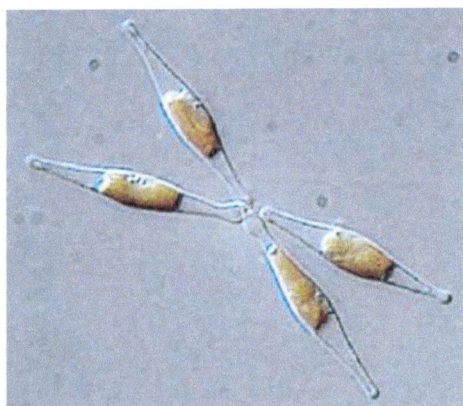

Fig. (21). *Phaeodactylum tricornutum* Alessandra de Martino and Chris Bowler,; Creative Commons Attribution 4.0 license ; https://commons.wikimedia.org/wiki/File:Phaeodactylum_tricornutum.png.

Porosira glacialis

Global distribution: Cold water habitats in the northern and southern hemispheres.

Ecology: It is pelagic or attached to various components of marine life.

Antidiabetics and their mechanisms of action:

The crude extracts of this species cultivated at high temperature–high light show antidiabetic activity by inhibiting the protein tyrosine phosphatase 1B (PTP1B) enzyme [8].

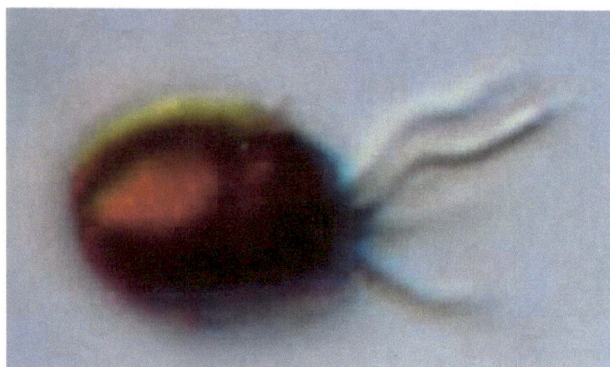

Fig. (22). *Isochrysis galbana.* F. Jouenne; Creative Commons Attribution-Share Alike 3.0 Netherlands license. https://commons.wikimedia.org/wiki/File:Isochrysis_galbana.jpg.

Antidiabetic Haptophyte Alga

Isochrysis galbana

Global distribution: Coastal, Atlantic, Mediterranean, and Pacific. (Fig. **22**)

Ecology: It is a brackish water species.

Antidiabetics and their mechanisms of action: Docosahexaenoic acid (DHA) (Fig. **23**) and eicosapentaenoic acid (EPA) derived from this species have been reported to regulate glucose metabolism in diabetic rats [25].

Fig. (23). Docosahexaenoic acid.

Antidiabetic Green Algae

The microalgae of chlorophyta generally occur in freshwater. However, many of them are known to be present in euryhaline conditions, including brackish waters and estuaries. Such species with medicinal importance are given below:

Auxenochlorella pyrenoidosa (= Chlorella pyrenoidosa)

Antidiabetics and their mechanisms of action: The extracts of this species have been reported to possess α-amylase and α-glucosidase inhibitory activities [8].

Auxenochlorella protothecoides (= *Chlorella protothecoides*)

Antidiabetics and their mechanisms of action: The pigments present in this species, *viz.* linoleic acid, arachidonic acid, astaxanthin, and lutein eicosapentaenoic acid, have been reported to prevent the formation of both endogenous and exogenous AGEs (Advanced Glycation End products), the inhibition of which is yet another mode of diabetes treatment. The high concentrations of these pigments are beneficial food ingredients and possible preventive agents for diabetic retinopathy patients. The biotechnology company Solazyme in the United States possesses a patent for the use of this species to treat patients with impaired glucose tolerance and diabetes (US 8747834 B2) [8].

Chlorella vulgaris

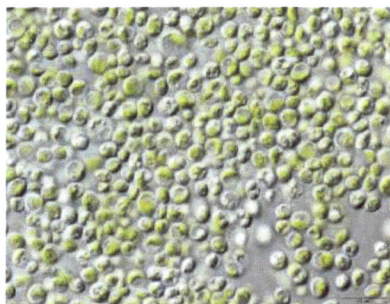

Fig. (24). *Chlorella vulgaris* ja:User:NEON / User:NEON_ja ; Creative Commons Attribution-Share Alike 3.0 Unported license. https://commons.wikimedia.org/wiki/ File:Chlorella_vulgaris_NIES2170.jpg

Antidiabetics and their mechanisms of action: The presence of carotenoid pigments, *viz.* linoleic acid, arachidonic acid, and eicosapentaenoic acid, has been reported to prevent the formation of AGE (Advanced Glycation End products) [8] (Fig. **24**).

Chlorella zofingiensis

Antidiabetics and their mechanisms of action: The extracts of this species containing rich astaxanthin possessed stronger antiglycative effects by inhibiting the formation of AGE. It is suggested that this microalga can be a possible preventive agent for diabetic patients [8]. Its omega-3 fatty acid, eicosapentaenoic acid, is a potent preventive agent in patients with diabetic retinopathy [6].

Antidiabetic Ochrophyte Algae

Nannochlorpsis gaditana

Global distribution: It is a common species of Spain.

Ecology: It is often found in both fresh and marine environments.

Antidiabetics and their mechanisms of action: The crude extracts of this species have been reported to possess antidiabetic properties [26].

Nannochloropsis oculata

Global distribution: It is widely distributed in all oceans.

Ecology: It is found in both marine and freshwater.

Antidiabetics and their mechanisms of action: The ethyl acetate crude extracts of this species have shown both α-amylase and α-glucosidase inhibitory activities with IC50 values of 121.96 and 178.53 µg/mL, respectively, and percentage values of 78.52 and 80.42%, respectively [18].

Nannochlorpsis sp. (Fig. **25**)

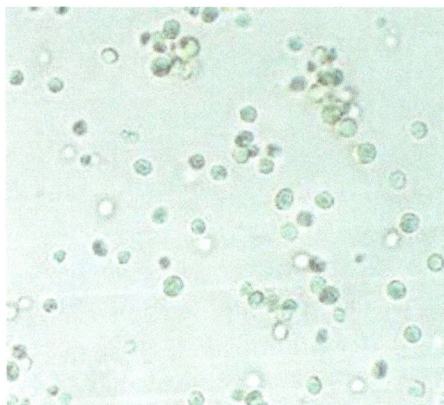

Fig.   (25).   *Nannochlorpsis*   sp.   Inks002,   public   domain   ;   https://commons. wikimedia.org/wiki/File:15_3klein2.jpg.

Antidiabetics and their mechanisms of action: The presence of both docosahexaenoic acid and eicosapentaenoic acid in this species has been reported to regulate glucose metabolism [25]. The administration of these algae was found to increase the production of low-density lipoproteins and decrease high-density lipoproteins in healthy and diabetic rats [8].

Antidiabetic Red Alga

Porphyridium sp. (Fig. **26**)

Neobodo, Creative Commons Attribution-Share Alike 4.0 International license.

Global distribution: It has a worldwide distribution.

Ecology: It occurs in both marine and freshwater habitats.

Antidiabetics and their mechanisms of action: The methanol crude extract of this species shows α-glucosidase inhibitory activity with a percentage value of 12.63 [18].

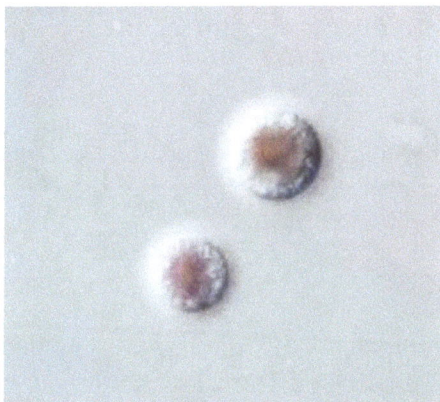

Fig. (26). *Porphyridium* **sp**. https://en.m.wikipedia.org/wiki/File:Porphyridium_purpureum1.jpg

Antidiabetic Marine Bacteria

Several species of marine bacteria have shown antidiabetic properties; such strains are listed below:

Paenibacillus sp. *TKU042*:

Antidiabetics and their mechanisms of action: The fermented demineralized crab shell powder (Fig. **27**) and shrimp shell powder using Paenibacillus sp. TKU042 have been reported to possess antidiabetic activity by inhibiting α-glucosidase [6].

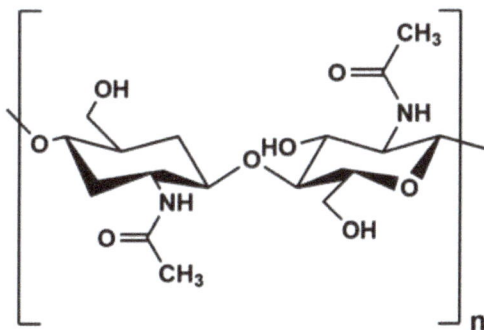

Fig. (27). Demineralized crab shell powder.

Streptomyces corchorusii

Antidiabetics and their mechanisms of action: The extracts of this bacterial species showed α-amylase inhibitory activity [8].

Streptomyces corchorusii **subsp.** *rhodomarinus*

Antidiabetics and their mechanisms of action: The extracts of this bacterial species showed α-amylase inhibitory activity [8].

Streptomyces **sp.**

Antidiabetics and their mechanisms of action: The compounds pyrostatins A and B derived from this species showed specific inhibitory activity against N-acetyl glucosaminidase [8].

Streptomyces strain, SSC21

Antidiabetics and their mechanisms of action: The terpenoid compound *viz.* suncheonoside A Fig. (**28**) derived from this strain showed antidiabetic activity (IC50,=10 uM) by producing adiponectin [27].

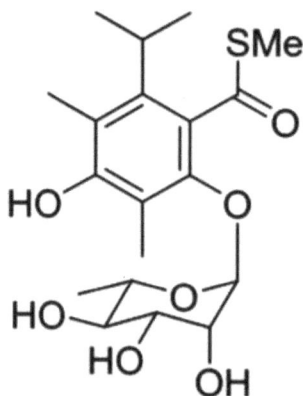

Fig. (28). Suncheonoside A.

Other Strains of Bacteria

The β-glucosidase inhibition activity of several strains of marine bacteria is listed in Table **1**.

Table 1. β-Glucosidase inhibition activity of marine bacterial strains [28].

Strain	Closest Match	β-Glucosidase Inhibition (mm)
GDA11	*Streptomyces coelicoflavus*	4
GDB16	*Bacillus* sp.	5
GDN4	*Vibrio communis*	5
GPB10	*Pseudochrobactrum* sp.	4
GPB13	*Bacillus aryabhattai*	4
GPB20	*Bacillus aryabhattai*	4
GPB21	*Staphylococcus gallinarum*	4
GPB8	*Staphylococcus gallinarum*	4
GPB9	*Bacillus subtilis* subsp. *spizizenii*	4
SD1-1	*Stenotrophomonas rhizophila*	4
SD1-13	*Arthrobacter koreensis*	4
SD1-14(1)	*Planomicrobium okeanokoites*	6
SD1-17	*Dietzia maris*	4
SD1-18	*Chryseomicrobium* sp.	4
SD1-20(1)	*Sphingobacterium* sp.	4
SD1-23	*Exiguobacterium marinum*	4
SD1-25	*Stenotrophomonas rhizophila*	4
SD1-3	*Arthrobacter koreensis*	4
SD1-6(1)	*Advenella kashmirensis*	5
SD1-8	*Microbacterium oleivorans*	5
SD2-1	*Bacillus flexus*	4
SD2-15(1)	*Exiguobacterium* sp.	4
SD2-17	*Bacillus oceanisediminis*	4
SD2-18	*Microbacterium esteraromaticum*	4
SD2-2(1)	*Bacillus oceanisediminis*	5
SD2-2(2)	*Bacillus siamensis*	5
SD2-20	*Bacillus methylotrophicus*	4
SD2-22	*Bacillus subtilis* subsp. *inaquosorum*	5
SD2-24	*Psychrobacter maritimus*	4
SD2-3(2)	*Bacillus flexus*	6
SD2-5	*Bacillus* sp.	4
SD2-6(1)	*Arthrobacter koreensis*	5
SD2-7(2)	*Bacillus stratosphericus*	5

(Table 1) cont.....

Strain	Closest Match	β-Glucosidase Inhibition (mm)
SP2A6	*Streptomyces rangoonensis*	4
SP2B11	*Planococcus rifitoensis*	5
SP2B12	*Bacillus stratosphericus*	5
SP2B20	*Bacillus amyloliquefaciens subsp. amyloliquefaciens*	6
SP2B3	*Halomonas sulfidaeris*	6
SP2B5	*Bacillus tequilensis*	4
SP2B6	*Bacillus* sp.	6
SP2B9	*Leucobacter chromiiresistens*	6

Antidiabetic Marine Fungi

Several species of marine fungi demonstrate antidiabetic properties. However, the species *Aspergillus* has been extensively studied in this aspect.

Aspergillus aculeatus CR1323-04

Antidiabetics and their mechanisms of action: The compound aspergillusol A derived from this fungus showed α-glucosidase inhibitory activity [29].

Aspergillus flavipes HN4-13

Antidiabetics and their mechanisms of action: Three butenolide derivatives, *viz.* flavipesolide A-C (Figs. **29-31**), derived from this fungus, acted as potent α-glucosidase inhibitors [6].

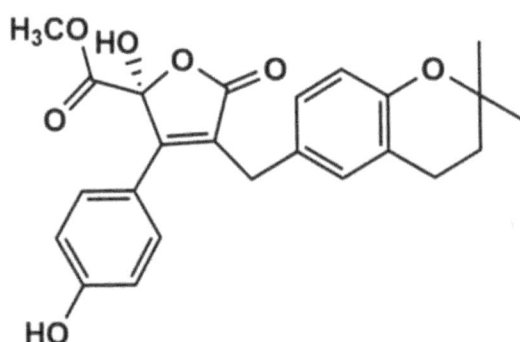

Fig. (29). Flavipesolide A.

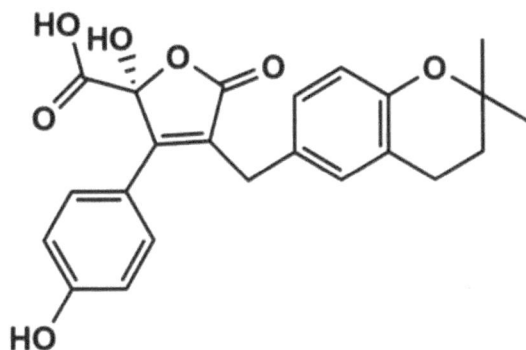

Fig. (30). Flavipesolide B.

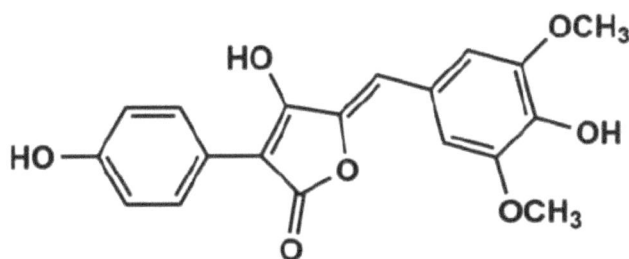

Fig. (31). Lavipesolide C.

Aspergillus terreus

The different strains of this fungal species have been reported to be associated with different components of marine life and possess several bioactive compounds with antidiabetic activity, as shown in Tables **2** and **3**.

Table 2. Strains of *Aspergillus terreus* and their associated marine life [30].

Strain of Aspergillus Terreus	Associated Marine Life
SCSGAF0162	Gorgonian *Echinogorgia aurantiaca* Tissue
SCSIO 41008	Marine Sponge, *Callyspongia* sp.
Not Identified	Red Alga *Halymenia acuminata*
BDKU 1164	Marine Sponge, *Haliclona* sp.
ML-44	Pacific Oyster's Gut
MXH-23	Unidentified sponge
GX7-3B	Mangrove Plant, *Bruguiera gymnoihiza*
TJ403-A1	Soft Coral *Sarcophyton subviride*
H010	Mangrove Plant, *Kandelia obovata*

(Table 2) cont.....

Strain of Aspergillus Terreus	Associated Marine Life
GZU-31-1	Mollluscan Sea Slug, *Onchidium struma*
Not identified	Marine Sponge *Phakellia. fusca*
Not identified	Red Alga, *Laurencia ceylanica*
EN-539	Red Alga *Laurencia okamurai*

Antidiabetics and their mechanisms of action: The compounds derived from this fungal species have shown antidiabetic activity by inhibiting α-glucosidase. Such compounds and their IC50 values are given in the following Table **3**.

Table 3. Bioactive compounds of *Aspergillus terreus* with α-glucosidase inhibitory activity [30].

Compound	Chemistry	IC50, μM
Butyrolactone I (Fig. **32**)	Butenolide	3.9
Aspernolide E (Fig. **33**)	-do-	8.1
Butyrolactone VII (Fig. **34**)	-do-	1.4
(-)-Asperteretal D (Fig. **35**)	Butenolide derivative	10
Asperteretal D	-do-	8.7
Asperteretal E (Fig. **36**)	-do-	13.4
Flavipesolide B (Fig. **37**)	-do-	10.3
Flavipesolide C (Fig. **38**)	-do-	7.6
5-[(3,4-dihydro-2,2-dimethyl-2H-1-benzopyran-6-yl)-methyl]-3-hydroxy-4-(4-hydroxyphenyl)-2(5H)-furanone (Fig. **39**)	y-buterolactone	11.7
Aspernolide A (Fig. **40**)	Butenolide	47.3
Amauromine (Fig. **41**)	Alkaloid	0.3
Austalide N (Fig. **42**)	Meroterpenoid	0.7
Prenylated diketopiperazine	Alkaloid	0.4
Versicolactone G (Fig. **43**)	Butenolide	104.8
Cowabenzophenone A (Fig. **44**)	tetracyclo[7.3.3.33, 11.03,7]tetradecane-2,12,14-trione derivative	7.8
Kodaistatins A–D	Natural products	80-130*

* values in nM

Its Kodaistatin A and C (Fig. **45**) suppressed glucose-6-phosphatase activity with IC50 values of 0.08 and 0.13 µM, respectively [30]. Further, its terrelumamide A and B (Figs. **46** and **47**) were found to improve insulin sensitivity by accelerating the production of adiponectin in the hBM-MSCs adipogenesis model. Its butanolide derivatives asperteretal D and E from Aspergillus terreus associated with the marine sponge Phakellia fusca displayed α-glucosidase inhibitory activity [27].

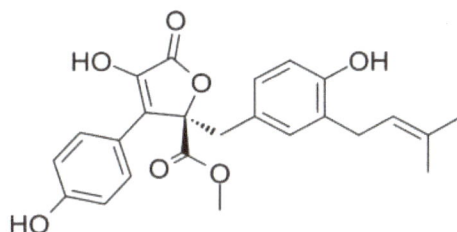

Fig. (32). Butyrolactone I.

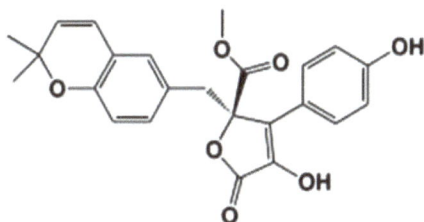

Fig. (33). Aspernolide E.

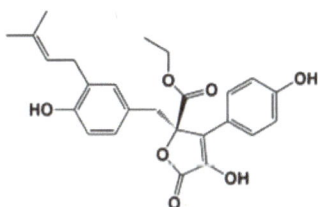

Fig. (34). Butyrolactone VII.

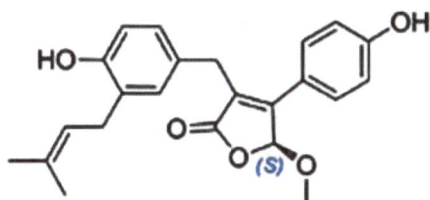

Fig. (35). Asperteretal D.

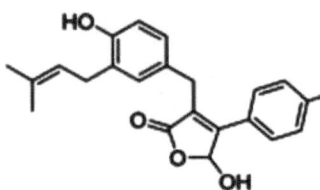

Fig. (36). Asperteretal E.

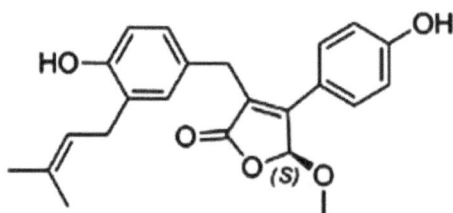

Fig. (37). Flavipesolide B (R=CH3).

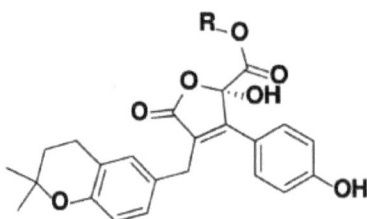

Fig. (38). Flavipesolide C (R=H).

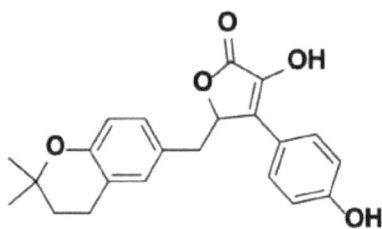

Fig. (39). 5-[(3,4-dihydro-2,2-dimethyl-2H-1-benzopyran-6-yl)-methyl]-3-hydroxy-4 -(4-hydroxyphenyl)-2(5H)-furanone.

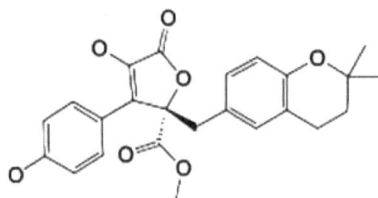

Fig. (40). Aspernolide A.

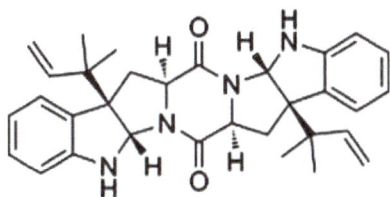

Fig. (41). Amauromine.

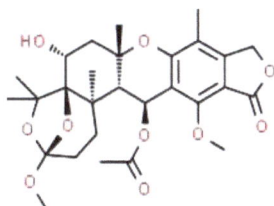

Fig. (42). Austalide N.

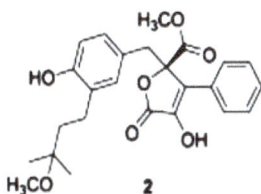

Fig. (43). Versicolactone G.

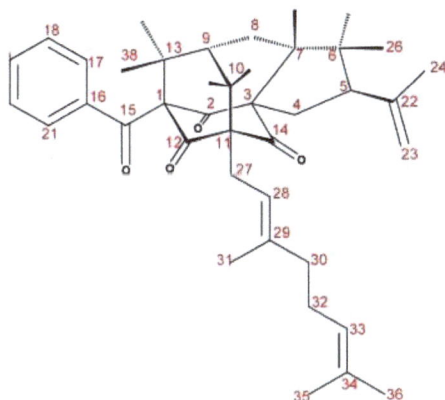

Fig. (44). Cowabenzophenone A.

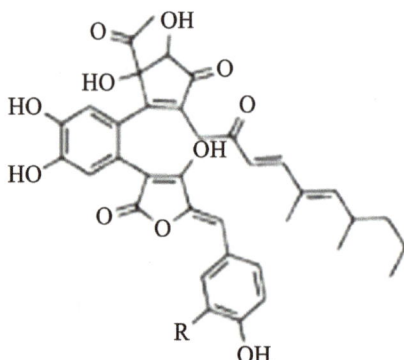

Fig. (45). Kodaistatin A (R=H).

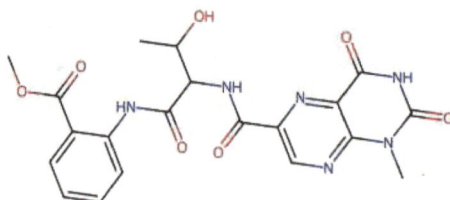

Fig. (46). Terrelumamide A.

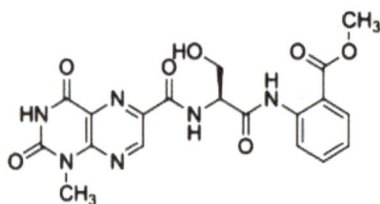

Fig. (47). Terrelumamide B.

Aspergillus unguis isolate SP51-EGY

Antidiabetics and their mechanisms of action: Streptozotocin-induced diabetic mice were treated with the shake filtrate and mycelial extracts of this fungal species, and both the extracts displayed maximum *in vitro* α-glucosidase inhibitory activity. At 50 and 100 mg/kg, *in vivo*, treatment with shake mycelia potently reduced the blood glucose level by 27% and 54%, respectively. On the other hand, at 125 and 250 mg/kg, only the shake filtrate significantly reduced the blood glucose level by 49% and 70%, respectively [31].

Aspergillus sp.

Antidiabetics and their mechanisms of action: The butenolide derivatives phenyl- and benzyl disubstituted γ-butenolide (Figs. **48** and **49**) derived from a coral-associated Aspergillus sp. displayed α-glucosidase inhibition activity [6].

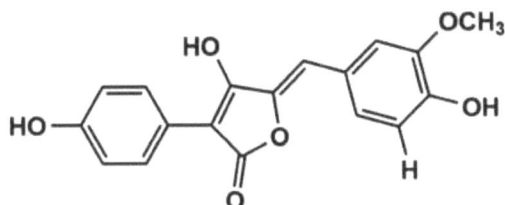

Fig. (48). Phenyl- disubstituted γ-butenolide.

Cosmospora sp.

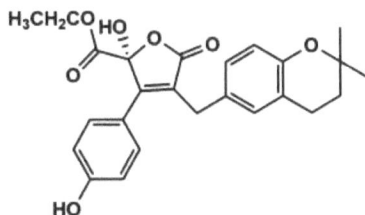

Fig. (49). Benzyldisubstituted γ-butenolide.

Antidiabetics and their mechanisms of action: The fungal metabolite, aquastatin A (Fig. **50**), derived from this fungal species displayed significant inhibitory activity against PTP1B with an IC50 value of 0.2 μM [24].

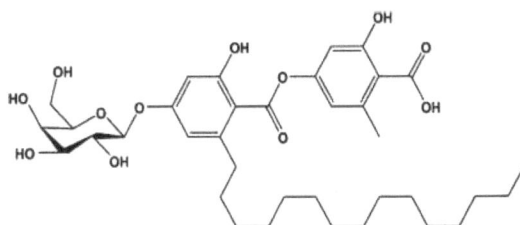

Fig. (50). Aquastatin A.

Eurotium spp. and *Penicillium* sp.

Antidiabetics and their mechanisms of action: The compounds fructigenine A, cyclopenol, echinulin, flavoglaucin, and viridicatol (Figs. **51-55**) derived from the strains of these fungal species show PTP1B inhibitory activity with IC50 values of 10.7, 30.0, 29.4, 13.4, and 64.0 μM, respectively [24].

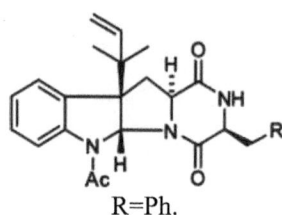

R=Ph.

Fig. (51). Fructigenine A.

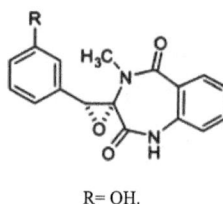

R= OH.

Fig. (52). Cyclopenol.

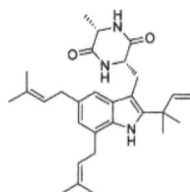

Fig. (53). Echinulin.

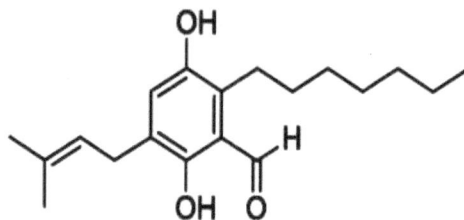

Fig. (54). Flavoglucin.

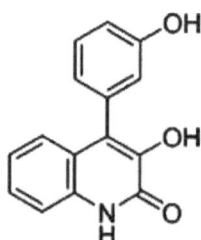

Fig. (55). Viridicatol.

Antidiabetics and their mechanisms of action: Polysaccharides, chromium (III) complexesderived from this fungal species, have been reported to increase the glucose tolerance capacity in the experimental diabetic mice subjected to oral glucose tolerance test [6].

Penicillium sp. JF-55

Antidiabetics and their mechanisms of action: The methylethylketone derivative isolated from this fungal strain displayed hypoglycemic effects by inhibiting PTP. Further, this fungal strain yielded the compounds penstyrylpyrone, anhydrofulvic acid, and citromycetin (Figs. **56-58**). Among these compounds, penstyrylpyrone and anhydrofulvic acid displayed PTP inhibitory activity with IC50 values of 5.28 and 1.90 µM, respectively [3].

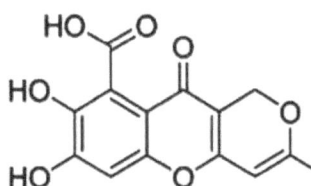

Fig. (56). Penstyrylpyrone.

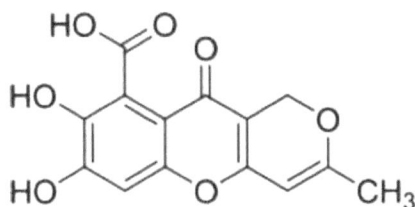

Fig. (57). Anhydrofulvic acid.

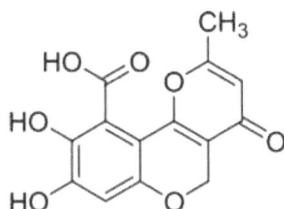

Fig. (58). Citromycetin.

Penicillium sp. SF-6013

Antidiabetics and their mechanisms of action: The compound terrelumamide A, derived from this fungal strain, showed antidiabetic activity with an IC50 value of 37 uM besides enhancing the production of adiponectin, a hormone that helps with insulin sensitivity [27].

Trichoderma sp. strain 307

Antidiabetics and their mechanisms of action: This endophytic fungal strain has been found to be associated with the mangrove plant Clerodendrum inerme. A polyketide depsidone botryorhodine H (Fig. **59**) isolated from this fungus showed α-glucosidase inhibiting activity [27].

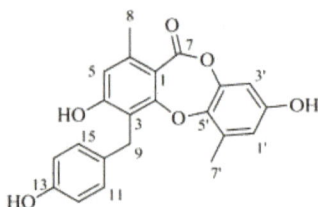

Fig. (59). Botryorhodine H.

Antidiabetic Antarctic Lichen

Lecidella carpathica

Fig. (60). *Lecidella carpathica.*

Ed Uebel , Creative Commons Attribution-Share Alike 3.0 Unported license.

https://en.m.wikipedia.org/wiki/File:Lecidella_carpathica.jpg

Global distribution: North and South America, Africa, Asia, and Europe.

Ecology: It inhabits non-calciferous rocks and rarely wood or bark (Fig. **60**).

Antidiabetics and their mechanisms of action: The MeOH extract of this species yields hopane-6α,22-diol, brialmontin 1, atraric acid, and two aromatic metabolites (Figs. **61-65**). Among these compounds, hopane-6α,22-diol, brialmontin 1, and atraric acid displayed PTP1B inhibitory activity with IC50 values of 3.7, 14.0, and 51.5 μM, respectively [32].

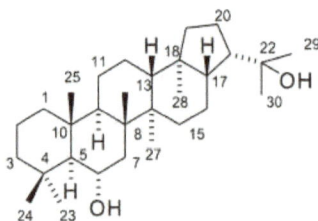

Fig. (61). Hopane-6α,22-diol.

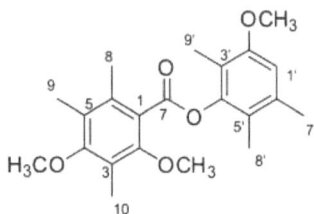

Fig. (62). Brialmontin 1.

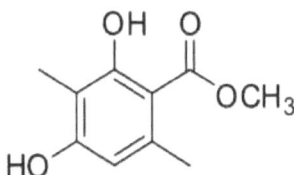

Fig. (63). Atraric acid.

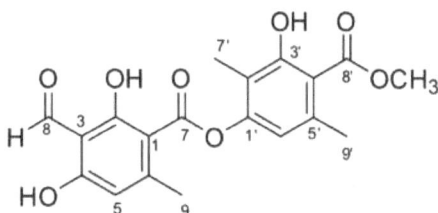

Fig. (64). Aromatic metabolite 1.

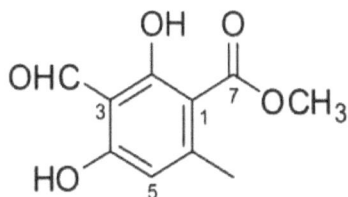

Fig, (65). Aromatic metabolite 2.

CONCLUSION

Among the marine microorganisms, marine bacteria and fungi have attracted the attention of scientists in the discovery of drugs for infectious diseases and cancer. On the other hand, many other pharmacologically active compounds, including antidiabetic compounds of these microbes, are being largely overlooked. Further, among the different classes of microalgae, only the cyanobacteria (blue-green algae) are in the active interest of the researchers owing to their rich contribution to the development of cyanotherapeutics, an emerging field for future drug discovery. This calls for intensive research on the chemical and pharmacological properties of other bioactive compounds of marine bacteria and fungi and on the identification of new and potential bioactive compounds from the other neglected groups of marine microalgae, such as diatoms and dinoflagellates.

CHAPTER 4

Antidiabetic Potential of Marine Plants

Abstract: The antidiabetic potentials of the different classes of marine plants, such as macroalgae *viz.* green algae, brown algae, and red algae, as well as seagrasses and mangrove plants are dealt with in this chapter. The different chemical classes of secondary metabolites derived from these marine plants and their mechanisms of action in antidiabetic activities are also given.

Keywords: Antidiabetic potentials, α-amylase inhibitory activity, α-glucosidase inhibitory activity, Brown algae, DPP-IV inhibitory activity, Green algae, Macroalgae, Mangrove plants, PTP 1B inhibitory activity, Red algae, Sea grasses.

INTRODUCTION

Marine macroalgae or seaweeds are large visible plants that generally grow by attaching to rocks along the seashore. These algae possess nutritional applications, such as human food and animal feeds. Based on morphological characteristics, anatomical features, pigment content, nutrients, and chemical composition, these seaweeds are classified as Chlorophyta (green algae), Phaeophyceae (brown algae), and Rhodophyta (red algae). The bioactive compounds or secondary metabolites isolated from the different types of these seaweeds belong to chemical classes such as alkaloids, terpenoids, flavonoids, steroids, and phenolsand have received greater attention in recent years due to their unique and diverse therapeutic properties, including antidiabetic and antioxidant. Among these compounds, the alkaloids are known to possess cytotoxic activity, terpenoids exhibit a wide spectrum of anti-tumor activities, phenolics exert significant antioxidant activities, steroids possess antimicrobial and cardiotonic properties, tannins serve as antioxidant, antiviral, antibacterial, antiulcer, and cytotoxic agents, and flavonoids possess antioxidant, antimicrobial, and spasmolytic activities. The other miscellaneous compounds of these seaweeds, like saponins, are found to be useful in hyperglycemia and hypercholesterolemia and as anti-inflammatory, anticancer, and weight-loss drugs [18] The presently available treatment regimens for type 2 DM have been found to possess adverse side effects, and there is a great need to search for effective and side effect-free drugs that can help maintain the blood glucose level and complications in type 2 DM patients. Even though herbal medicines have been focused on by most of the researchers, none of these medicines have yielded a fully beneficial effect on treating patients with type 2 diabetes mellitus. The seaweeds, with their promising antidiabetic compounds, offer vast scope in the development of new and safe diabetic drugs. The antidiabetic potentials of the different classes of seaweeds are given below.

MARINE MACROALGAE (SEAWEEDS)

Green Algae

Auxenochlorella pyrenoidosa (= Chlorella pyrenoidosa)

Global distribution: It occurs worldwide.

Ecology: It is a freshwater species and is occasionally seen in marine environments.

Antidiabetics and their mechanisms of action: The extracts of this species have been reported to suppress hyperglycaemic conditions by inhibiting α-glucosidase and α-amylase enzymes in type 2 diabetes mellitus patients [33].

Capsosiphon fulvescens

Global distribution: Southwest coast of South Korea.

Ecology: It is known to reside in the upper portions of the intertidal coastal sediments and rocky shores.

Antidiabetics and their mechanisms of action: A total of three glycolipids *viz.* capsofulvesin A ((2S)-1-O-(6Z,9Z,12Z,15Z-octadecatetraenoyl)-2- O-(4Z,7Z,10Z,13Z-hexadecatetraenoyl)-3-O-β-D-galactopyranosyl glycerol), capsofulvesin B ((2S)-l-O- (9Z,12Z,15Z-octadecatrienoyl)-2-O-(10Z-13Z-hexadecadienoyl)-3-O-β-D-galactopyranosyl glycerol), and capsofulvesin C ((2S)-1-O-(6Z,9Z,12Z,15Z-octadecatetraenoyl)-3-O-β-D-galacatopyranosyl glycerol), as well as chalinasterol (Figs. **1-4**), have been isolated from this algal species. Of them, capsofulvesin A, B, and sterol chalinasterol exhibit antidiabetic activities by inhibiting aldose reductase with IC50 values of 52.5, 101.9, and 345.3 µM, respectively [23 - 33].

Caulerpa racemosa

Global distribution: Temperate and tropical seas; Eastern Mediterranean Sea.

Ecology: It dwells in shallow seas.

Antidiabetics and their mechanisms of action: Its acetone crude extract showed α-amylase inhibitory activity with an ED50 value of 0.09 mg/ml [23]. At 100 and 200 mg/kg concentrations, the ethanolic extract of this alga showed antidiabetic activity by significantly reducing the blood glucose levels in diabetic rats besides restoring the glucose uptake by hemidiaphragm and glucose transport by hepatic cells. Further, at 200 mg/kg, the ethanolic extract was found to restore the histoarchitecture of the pancreas (Fig. **5**) [34].

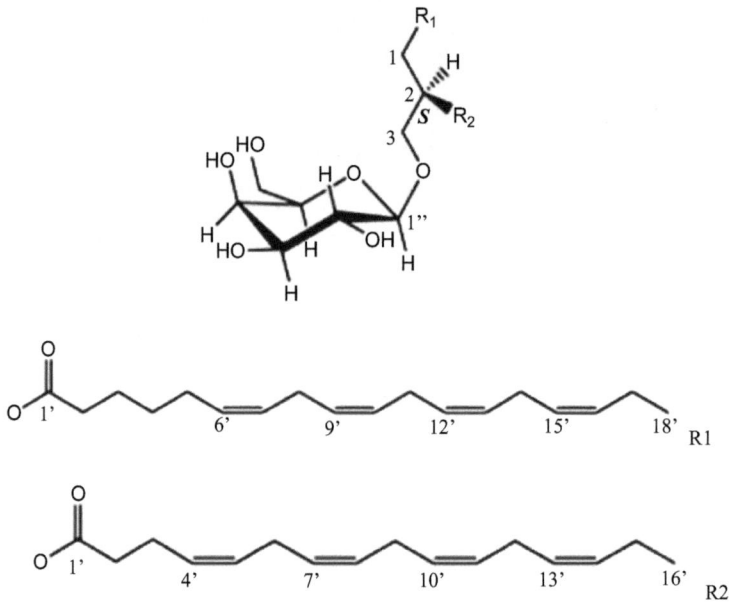

Fig. (1). Capsofulvesin A.

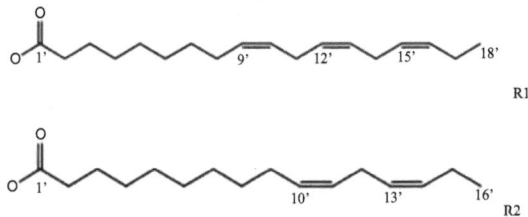

Fig. (2). Capsofulvesin B.

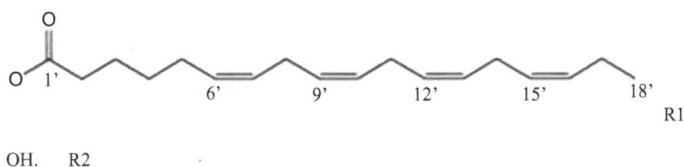

OH.   R2

Fig. (3). Capsofulvesin C.

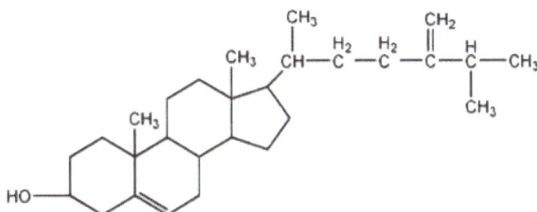

Fig. (4). Chalinasterol.

Fig. (5). *Caulerpa_racemosa*.Nhobgood Nick Hobgood; Creative Commons Attribution-Share Alike 3.0 Unported license. https://commons.wikimedia.org/wiki/File:Caulerpa_racemosa_algae.jpg.

Chaetomorpha aerea

Global distribution: It is wide spread worldwide and common in New Zealand.

Ecology: It inhabits both marine and freshwater biotopes. It usually grows in the lower parts of the intertidal or shallow rough waters where sand accumulates.

Antidiabetics and their mechanisms of action: The chloroform extract of this species has shown significant inhibitory activity on α-amylase enzyme with an IC50 value of 147.6 µg/ml [33]. Its extract has potent activity in managing diabetes by ably inhibiting the carbohydrate digestive enzymes [18].

Chaetomorpha antennina

Global distribution: It is commonly found distributed on the Visakhapatnam coast, Bay of Bengal, India.

Ecology: It is a coastal, benthic species. It also thrives in freshwater habitats.

Antidiabetics and their mechanisms of action: The methanolic extracts of this species have demonstrated antidiabetic activity by inhibiting alpha- amylase, alpha- glucosidase, and dipeptidyl peptidase-4 (DPPIV) with IC50 values of 525.8, 121.3, and 24.92 92 g mL1, respectively. Further, at a concentration of 250 mg kg1B.wt, this extract was found to reduce the fasting blood glucose level to 39.97% in streptozotocin-induced diabetic rats [35].

Chaetomorpha linum

Global distribution: Arctic to temperate waters in the south hemisphere; common in the Asia-Pacific Region.

Ecology: It is found attached to dead coral substrates and coral reefs in stagnant and shallow waters with little water exchange.

Antidiabetics and their mechanisms of action: A water-soluble polysaccharide *viz.* (sulfated rhamnogalactoarabinan (CHS2) isolated from this algal species displays significant antidiabetic activity by inhibiting the activity of human islet amyloid polypeptide (hIAPP) aggregation. It is also suggested that CHS2 can be a potential drug candidate in the development of a novel antidiabetic drug for type 2 diabetes mellitus treatment [36].

Chlorodesmis sp.

Global distribution: Tropical to subtropical waters.

Ecology: It is found from the upper sublittoral to 35m.

Antidiabetics and their mechanisms of action: At a concentration of 500 µg/ml, the methanol extract of this species has shown antidiabetic activity by inhibiting α-amylase enzyme by 72% with an IC50 value of 409 µg/ml [33].

Cladophora rupestris

Global distribution: North East Atlantic (from Arctic to Portugal and North Sea) and North West Atlantic (from Arctic to New Jersey).

Ecology: It dwells on the surface of rocks, in rock pools, hanging in 'ropes' in crevices, or forming undergrowth to macroalgae at all depths on the shore.

Antidiabetics and their mechanisms of action: At a concentration of 500 μg/ml, the methanol extract of this species shows a weak hypoglycemic effect by inhibiting the α-amylase with 14% (Fig. **6**) [33].

Fig. (6). *Cladophora_rupestris*-. Gabriele Kothe-Heinrich; Creative Commons Attribution-Share Alike 3.0 Unported license. https://commons.wikimedia.org/wiki/File:Cladophora_rupestris_Helgoland.jpg.

Codium adhaerens

Global distribution: Tropical to temperate areas; Atlantic, Indian, and Pacific Oceans.

Ecology: It is seen commonly growing on hard and rough substrates like vertical rocky walls and damaged corals in the middle and low intertidal zone.

Antidiabetics and their mechanisms of action: The chloroform and methanolic extracts of this species have been reported to possess antidiabetic activity by inhibiting the Protein Tyrosine Phosphatase 1B (PTP1B) enzyme and enhancing insulin sensitivity [33].

Derbesia marina

Global distribution: Throughout British and Irish coasts, western Scotland, and southwest Britain and Ireland.

Ecology: It is a subtidal species and is found attached to sponges, shells, and other algae down to depths of 20 m.

Antidiabetics and their mechanisms of action: The chloroform and methanolic extracts of this species have been reported to possess antidiabetic activity by inhibiting the Protein Tyrosine Phosphatase 1B (PTP1B) enzyme and enhancing insulin sensitivity [33].

Derbesia tenuissima

Global distribution: Temperate eastern Atlantic and Mediterranean regions.

Ecology: It is commonly seen in reef pools and marina pylons.

Antidiabetics and their mechanisms of action: At a concentration of 10 mg/ml, the crude extracts of this species show moderate α-amylase and potent α-glucosidase inhibitory activities with percentage values of 53.6 and 73.98%, respectively [33].

Halimeda macroloba

Global distribution: Tropics; Thai-Malay Peninsula and Florida Keys.

Ecology: It is usually seen in waters with a depth of less than a meter. It has been reported to grow best in fine and compact sand on shallow reef flats influenced by mild currents.

Antidiabetics and their mechanisms of action: The aqueous extract of this species displays α-glucosidase inhibitory activity with an IC50 value of 6.4 mg/mL [23]. At a 10 mg/ml concentration, the crude water extracts of this species exhibited DPP-4 enzyme inhibitory activity by 60.53% by stimulating Glucagon-like Peptide-1 (GLP-1) secretion (Fig. **7**) [33].

Oedogonium intermedium

Global distribution: South America: Brazil. Asia: China and India.

Ecology: It is known to prefer stagnant waters, such as small ponds, pools, marshes, lakes, and reservoirs. Occasionally, it may show its presence in euryhaline conditions.

Antidiabetics and their Mechanisms of action: At a concentration of 10 mg/ml, the crude extracts of this species have shown moderate α-amylase and potent α-glucosidase inhibitory activities with percentage values of 49.2 and 69.5%, respectively [33].

Rama rupestris (= Cladophora rupestris)

Global distribution: NE Atlantic (Arctic shores to Portugal and North Sea) and NW Atlantic.

Ecology: It dwells in rock pools and rock surfaces on the shore.

Antidiabetics and their mechanisms of action: The raw extracts of this algal species display α-glucosidase and α-amylase inhibitory activities [8]. It possesses potent activity to manage diabetes with the inhibition of the carbohydrate digestive enzymes [18].

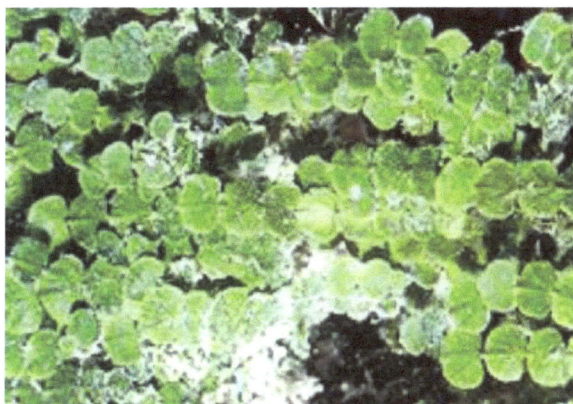

Fig. (7). *Halimeda* sp.; Dr. Robert Ricker, NOAA/NOS/ORR; public domain; https://commons.wikimedia.org/wiki/File:Halimeda_copiosa_at_10_meters_depth_in_shallow_cave.jpg.

Global distribution: It is commonly found in the seas around Australia.

Ecology: It is commonly found in open rocky coasts, estuaries with seagrass beds, mudflats, mussel-beds, and salt marshes.

Antidiabetics and their mechanisms of action: The ethanol extracts of this species showed α-glucosidase inhibitory activity at the concentrations of 250, 500, and 1000 µg/mL, and the percentage values recorded were 79.8, 86.9, and 96.3, respectively [37].

Ulva fenestrata

Global distribution: NE Atlantic and N Pacific.

Ecology: It is found on hard surfaces and harbors.

Antidiabetics and their mechanisms of action: A bioactive dark-red pigment *viz.* echinochrome A (7-ethyl-2,3,5,6,8-pentahydroxy-1,4-naphthoquinone) (Fig. **8**) (an active substance of the drug Histochrome, which has been approved for medicinal use in Russia against various diseases) of this algal species was found to improve glucose tolerance in seven-week-old diabetic mice when they were injected with Histochrome (0.3 mL/kg/day; EchA equivalent of 3 mg/kg/day) intraperitoneally for 12 weeks. Further, the potential reno-protective mechanism of EchA in the mouse model of type 2 DM may provide a new and potential therapeutic strategy for diabetic nephropathy [38].

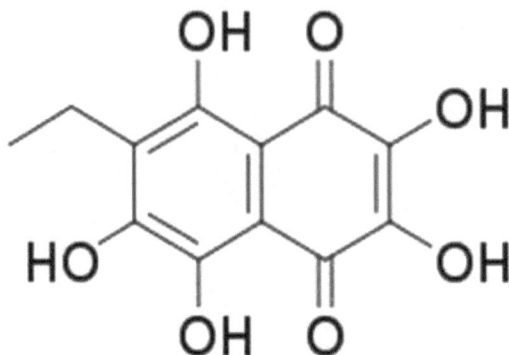

Fig. (8). Echinochrome A.

Ulva intestinalis (= Enteromorpha intestinalis)

Global distribution: It has more or less a worldwide distribution.

Ecology: It is seen in both fresh and saline waters, including ditches, pools, rockpools, and bedrock.

Antidiabetics and their mechanisms of action: The crude extracts of this species manage diabetes by inhibiting the enzymes of carbohydrate digestion [18].

Ulva lactuca (= Ulva fasciata)

Global distribution: Indo-Pacific and Atlantic.

Ecology: It commonly occurs in all intertidal areas, including brackish habitats.

Antidiabetics and their mechanisms of action: The extracts of this species have shown inhibitory activities of enzymes hexokinase, glucokinase, and glucose6-phosphatase in alloxan-induced diabetic rats. The percentage decrease of these enzymes was 56.9%, 79.9%, and 67.2%, respectively, in diabetic control (Fig. **9**) [39].

Fig. (9). *Ulva lactuca*; Kristian Peters; Creative Commons Attribution-Share Alike 3.0 Unported license. https://en.wikipedia.org/wiki/File:Ulva_lactuca.jpeg.

Ulva linza (= Enteromorpha linza)

Global distribution: Temperate North Atlantic and Mediterranean.

Ecology: This epiphytic species is seen on the rocky surfaces of the subtidal and low intertidal areas.

Antidiabetics and their mechanisms of action: The chloroform and methanolic extracts of this algal species were found to enhance insulin sensitivity by inhibiting the activities of the PTP1B enzyme [8].

Ulva ohnoi

Global distribution: Mainly in tropical to subtropical areas, including Pacific coastal areas.

Ecology: This shallow-water species is commonly found in tidal flats and estuaries.

Antidiabetics and their mechanisms of action: The crude extract of this species showed α-amylase inhibition by 41.7% and total α-glucosidase inhibition at 10 mg/mL [33].

Ulva prolifera (= Enteromorpha prolifera)

Global distribution: Tropical and subtropical waters of the Pacific Ocean.

Ecology: It is found widely distributed in the intertidal zone.

Antidiabetics and their mechanisms of action: The Sulfated Rhamnose Polysaccharides Chromium (III) (SRPC), a synthesized complex prepared from the sulfated rhamnose polysaccharide of this species, has shown increased glucose metabolism with hypoglycaemic effect, improved insulin resistance, and enhanced glucose tolerance in experimental T2DM mice induced with High-fat High-sucrose Diet (HFSD) (Fig. **10**) [6].

Fig. (10). *Ulva prolifera.* CathayanBoris; Creative Commons CC0 1.0 Universal Public Domain Dedication.; https://commons.wikimedia.org/wiki/File:Coastalprolifera.jpg.

Ulva reticulata

Global distribution: Malaysia, Indonesia, Japan, Philippines, the Indian Ocean, and the Red Sea.

Ecology: It is mostly seen in low intertidal to 2m deep water areas as an epiphyte.

Antidiabetics and their mechanisms of action: The aqueous extracts of this algal species exhibited antidiabetic effects by inhibiting α-glucosidase and α-amylase enzymes with percentage values of 76.02 and 89.1, respectively [6 - 33].

Ulva rigida

Ecology: It resides on coral reef flats and intertidal rocks. It is often abundant in areas of freshwater runoff, like near the mouths of streams.

Antidiabetics and their mechanisms of action: The ethanolic crude extracts of this algal species showed antidiabetic effects by inhibiting α-amylase and α-glucosidase with percentage values of 89.1 and 79.6, respectively [18]. Further, its raw extracts showed a reduction of plasma glucose levels in experimental rats (Fig. **11**) [8].

Fig. (11). *Ulva rigida-* Thesupermat; Creative Commons Attribution-Share Alike 2.5 Generic, 2.0 Generic and 1.0 Generic license. https://commons.wikimedia.org/wiki/File:Mar%C3%A9e_verte_-_Ulva_Armoricana_-_en_nord_Finist%C3%A8re_-_005.JPG.

Brown algae

Alaria sp.

Global distribution: It is often found in temperate seas.

Ecology: It is seen along the coasts.

Antidiabetics and their mechanisms of action: The phenolic extracts of this species show α-amylase inhibitory activity (Fig. **12**) [40].

Fig. (12). *Alaria* sp. Pierre-Louis Crouan (1798-1871) & Hippolyte-Marie Crouan (1802-1871); public domain ; https://commons.wikimedia.org/wiki/File:Alaria_esculenta_Crouan.jpg.

Ascophyllum nodosum

Global distribution: It is a major species of North Atlantic.

Ecology: It is commonly found on intertidal and subtidal rocks, especially in sheltered waters (Fig. **13**).

Antidiabetics and their mechanisms of action: A number of extracts of this algal species have been reported to exhibit antidiabetic effects in terms of α-amylase inhibition and α-glucosidase inhibition, and the IC50 values recorded for the various extracts are given in Table **1**.

Fig. (13). *Ascophyllum_nodosum*. Lairich Rig; Creative Commons Attribution-ShareAlike 2.0 license.; https://commons.wikimedia.org/wiki/File:Ascophyllum_nodosum_with_Polysiphonia_lanosa.jpg.

Table 1. Antidiabetic effects of extracts of Ascophyllum nodosum [23-41].

Extract	Antidiabetic Effect (s)	IC50 Value (s)
Water extract	α-amylase inhibition; α-glucosidase inhibition	1.34 µg phenolics; 0.24 µg phenolics
Ethanol and cold water extracts	α-amylase inhibition (water)	53.6 µg/mL
Phlorotannin-rich extract	α-amylase inhibition; α-glucosidase inhibition	~0.1 µg/mL GAE; ~20 µg/mL GAE*
Water ethanolic extract	α-glucosidase inhibition	77 µg/mL
Methanol extract	α-glucosidase and α-amylase inhibition	-----

Ascophyllum sp.

Antidiabetics and their mechanisms of action:

The extracts of this algal species exhibited significant α-amylase inhibitory activity with IC50 values of ~0.1 µg/ml GAE. On the other hand, these extracts weakly inhibited α-glucosidase, the key enzyme involved in blood glucose regulation and starch digestion, with IC50s of ~20 µg/ml GAE [40]. On the other hand, the extract of this species with rich phlorotannin showed weak activity against α-glucosidase (IC50 of 20 µg mL−1 GAE) [6].

Choonospora minima

Global distribution: It is widely distributed in the North Atlantic and Eastern Atlantic regions.

Ecology: It is known to dominate the intertidal zone.

Antidiabetics and their mechanisms of action: The extracts of this algal species displayed inhibitory activity against α-amylase with an IC50 value of 17.9 µg mL, and the values of percentage inhibition of the different extracts ranged from 89.6 to 3.7 [42].

Cystoseira moniliformis

Global distribution: Mediterranean and Northeast Atlantic.

Ecology: It is a coastal species.

Antidiabetics and their mechanisms of action: The extracts of this algal species have shown a significant type I antidiabetic effect against the blood glucose level of alloxan-induced hyperglycaemic albino mice (Fig. **14**) [43].

Fig. (14). *Cystoseira* sp.; Esculapio; Creative Commons Attribution-Share Alike 3.0 Unported license.; https://commons.wikimedia.org/wiki/File:Cystoseira_0001.JPG.

Dictyopteris divaricata

Global distribution: Commonly distributed in Pacific waters.

Ecology: It commonly inhabits littoral and sublittoral rock zones.

Antidiabetics and their mechanisms of action: The chloroform: methanol (1:2) extracts of this species exhibited alpha-glucosidase inhibitory effects with the calculated inhibition percentage of 62.78% at 79.6 µg/ml (Fig. **15**) [44].

Fig. (15). *Dictyopteris* sp.; B.navez; Creative Commons Attribution-Share Alike 3.0 Unported, 2.5 Generic, 2.0 Generic and 1.0 Generic license.; https://commons.wikimedia.org/wiki/File:Dictyopteris_membranacea_herbarium_item.jpg.

Dictyopteris hoytii

Global distribution: It is widely distributed in tropical, subtropical, and temperate regions.

Ecology: It is seen commonly in seamounts and knolls.

Antidiabetics and their mechanisms of action: The methanolic extract of this species has yielded compounds such as fucosterol (Fig. **16**), diethyl-2 bromobenzene 1,4-dioate (Fig. **17**), ethyl methyl 2-bromobenzene 1,4-dioate (Fig. **18**), n-hexadecanoic acid (Fig. **19**), methyl ester, cerotic acid (Fig. **20**), β-sitosterol (Fig. **21**), 11-eicosenoic acid (Fig. **22**), and n-octacos-9-enoic acid. Among these compounds, n-octacos-9-enoic acid was found to be most active against the α-glucosidase enzyme with an IC50 value of 30.5 µM. On the other hand, fucosterol and diethyl-2-bromobenzene 1,4-dioate showed fairly good inhibitory activity with IC50 values of 289.4 and 234.2 µM, respectively, and ethyl methyl 2-bromobenzene 1,4-dioate, β-sitosterol, and cerotic acid displayed only moderate and low inhibitory activity against α-glucosidase [44].

(Kindly note that the following four images (17,18, 20, and 21) are interchanged as per the revised text above).

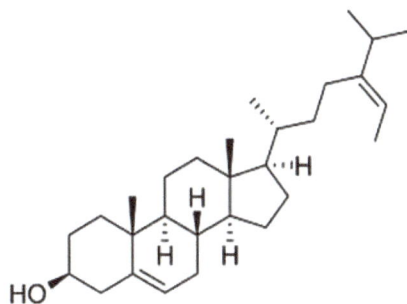

Fig. (16). Fucosterol.

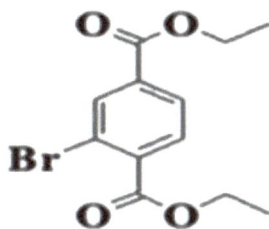

Fig. (17). Diethyl-2-bromobenzene 1,4-dioate.

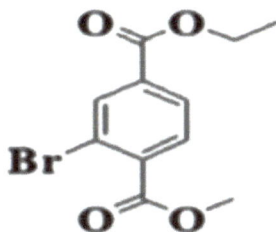

Fig. (18). Ethyl methyl 2-bromobenzene 1,4-dioate.

Fig. (19). n-hexadecanoic acid.

Fig. (20). Cerotic acid.

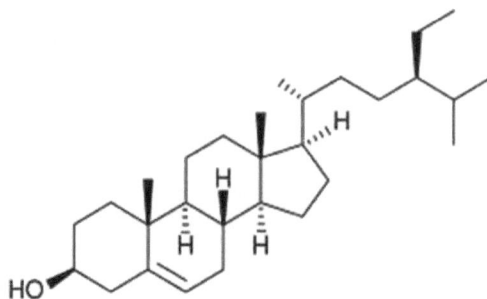

Fig. (21). β-sitosterol.

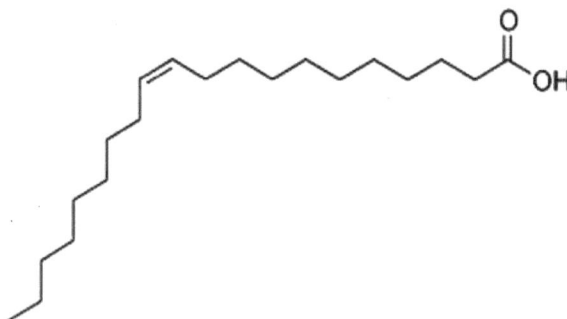

Fig. (22). 11-eicosenoic acid.

Dictyopteris prolifera

Global distribution: It is widely distributed in tropical, subtropical, and temperate regions.

Ecology: It is commonly seen in seamounts and knolls.

Antidiabetics and their mechanisms of action: The ethanolic extract of this species exhibited significant inhibitory activity against α-glucosidase with a percentage value of 99. 2 and an IC50 value of 16.66 μg/mL [44].

Dictyopteris undulata

Global distribution: Asia-Pacific: China, Japan, Taiwan, and Korea.

Ecology: It is found growing on rock, below the low-tide mark to subtidal areas.

Antidiabetics and their echanisms of action: The steroidal compounds (24S)-7a methoxy-stigmasta-5, 28-diene-3b, 24-diol and (24S)-7b-methoxy-stigmasta-5, 28-diene-3b, 24-diol derived from this species have been reported to show PTP 1B inhibitory activity [41]. The mixture of its compounds *viz.* dictyopterisin D, dictyopterisin E, and dictyopterisin I displayed potent PTP1B inhibitory activity with IC50 values of 1.88 - 3.47 μM [44].

Ecklonia cava

Global distribution: It is commonly found distributed on the southern coast of Korea and Japan in the Pacific Ocean.

Ecology: It grows atop rocks along shallow coastlines.

Antidiabetics and their mechanisms of action: The methanol and polyphenol extracts of this species displayed antidiabetic activities in experimental rats by reducing their plasma glucose levels. Further, its dieckol, 2-phloroeckol, and phlorofucofuroeckol-A (Figs. **23-25**) have shown α-glucosidase inhibitory activity; fucodiphloroethol G (Fig. **26**) exerted α-amylase inhibiting activity; 6,6-Bieckol (Fig. **27**), 2-phloroeckol and phlorofucofuroeckol-A had PTP 1B inhibitory activity; phlorofucofuroeckol-A exhibited aldose reductase inhibitory activity; and 7-Phloroeckol (Fig. **28**) and phlorofucofuroeckol-A demonstrated ACE inhibitory activity [41]. Five phloroglucinol derivatives yielded by this species were found to possess α-amylase and α-glucosidase inhibitory activities with IC50 values of 124.9 and 10.8 µmol L−1, respectively [6]. Further, its methanolic extracts have also shown the reduction of post-prandial blood glucose levels, which was partially due to the AMP-activated protein kinase/ACC and PI-3K/Akt cellular signal pathways [23]. Its minor phlorotannin derivative *viz.* 8,8'-bieckol (Fig. **29**) possessed α-glucosidase inhibitory activity, and it is suggested that this compound has the potential to replace acarbose as a novel α-glucosidase inhibitor [27].

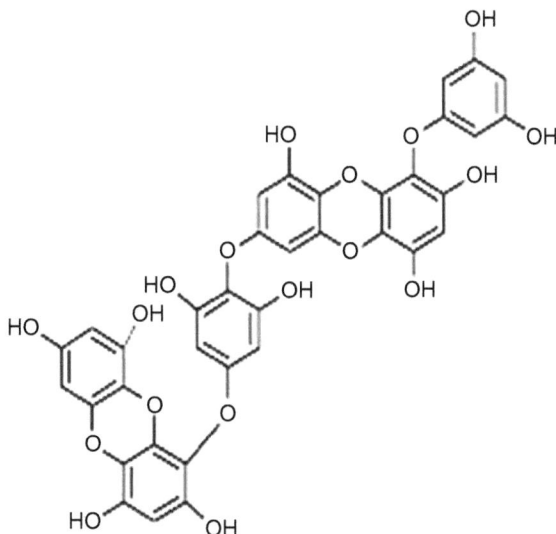

Fig. (23). Dieckol.

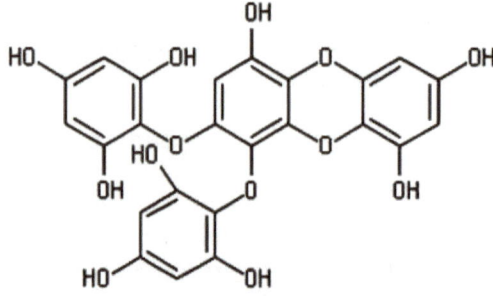

Fig. (24). 2-phloroeckol.

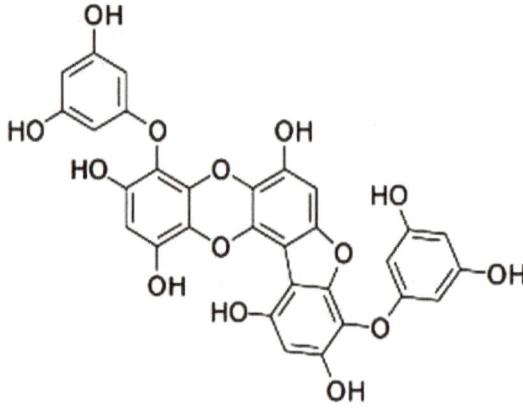

Fig. (25). Phlorofucofuroeckol-A.

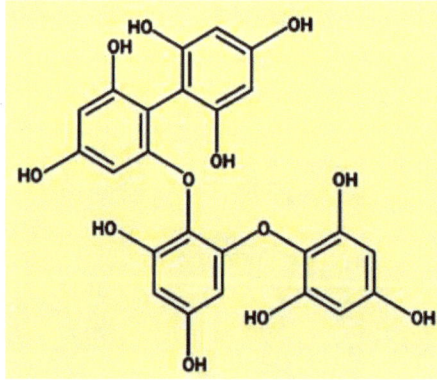

Fig. (26). Fucodiphloroethol G.

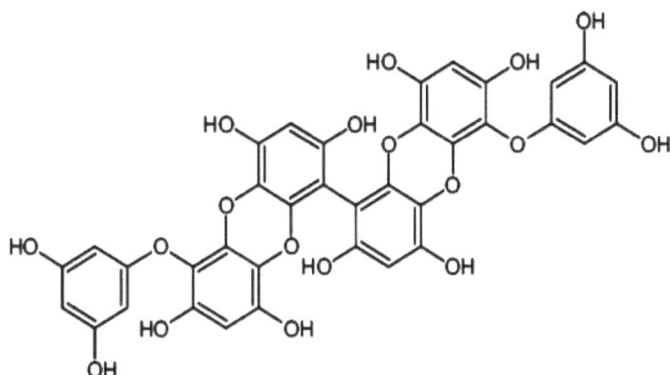

Fig. (27). 6,6-Bieckol.

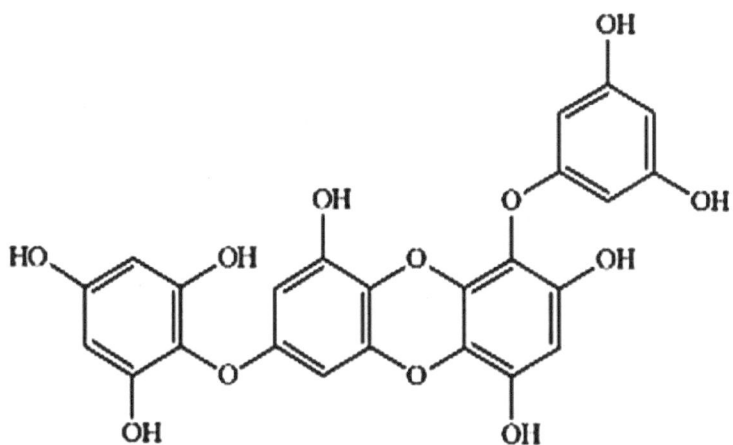

Fig. (28). 7-Phlorocckol.

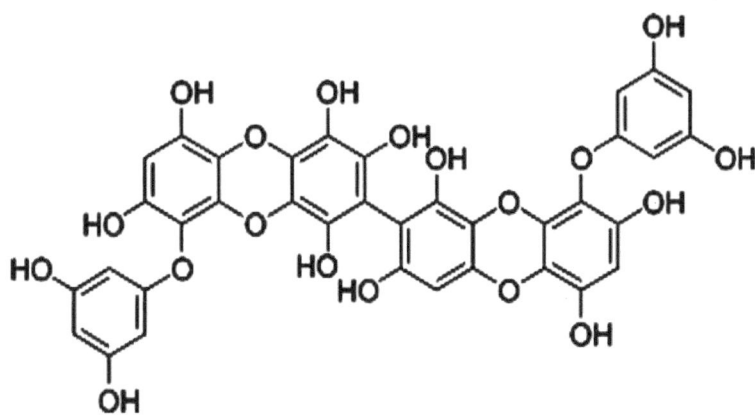

Fig. (29). 8,8'-bieckol.

Ecklonia maxima

Global distribution: Southern Atlantic: South Africa to northern Namibia.

Ecology: It dominates shallow waters up to a depth of 8 m in the offshore kelp forests (Fig. **30**).

Antidiabetics and their mechanisms of action: The phlorotannins, dibenzo 1,4-dioxin-2,4,7,9-tetraol   (Fig.   **31**)   (IC50   =   33.69   µM)   and hexahydroxyphenoxydibenzo-1,4-dioxin, derived from this species displayed significant α-glucosidase inhibition. Further, its eckol and phloroglucinol (Figs. **32**, **33**) displayed similar activity with IC50s of 11.2 and 19.9 µM, respectively [6 - 23].

Fig. (30). *Ecklonia_maxima.* Ossewa; Creative Commons Attribution-Share Alike 4.0 International license.; https://commons.wikimedia.org/wiki/File:Ecklonia_maxima,_Lambert%27s_Bay.jpg

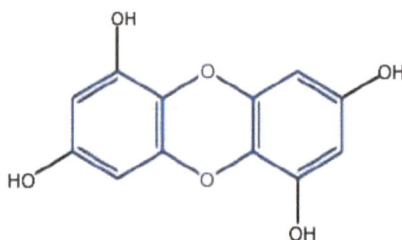

Fig. (31). Dibenzo [1,4] dioxine-2,4,7,9-tetraol.

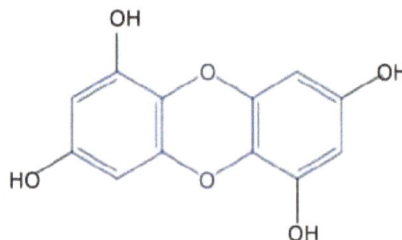

Fig. (32). Eckol.

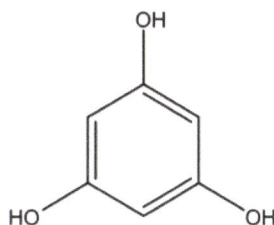

Fig. (33). Phloroglucinol.

Ecklonia stolonifera

Global distribution: Middle Pacific, around Korea and Japan (Fig. **34**).

Ecology: It is a subtidal species.

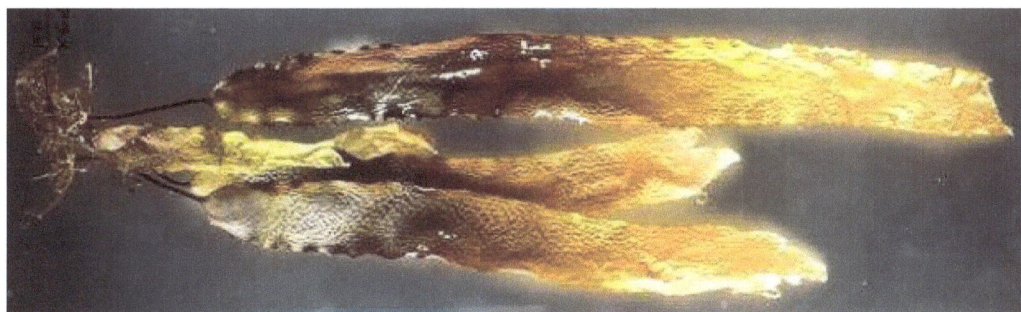

Fig. (34). *Ecklonia_stolonifera.* Daderot; Creative Commons CC0 1.0 Universal Public Domain Dedication.; https://commons.wikimedia.org/wiki/File:Ecklonia_stolonifera_-
_National_Museum_of_Nature_and_Science,_Tokyo_-_DSC07668.JPG

Antidiabetics and their mechanisms of action: The antidiabetic phlorotannins of this species, such as eckol, dieckol, phlorofucofuroeckol A, and phloroglucinol and its antidiabetic sterol *viz.* fucosterol with their modes of action are shown in Table **2**.

Table 2. Antidiabetic compounds of *Ecklonia stolonifera* and their mechanisms of action [41].

Antidiabetic Compound	Chemical Class	Mechanism(s) of Action
Eckol	Phlorotannin	PTP 1B and α-glucosidase inhibitor
Dieckol	-do-	α-amylase and ACE inhibitor
Phlorofucofuroeckol A	-do-	PTP 1B, α-glucosidase, α-amylase, and ACE inhibitor
Phloroglucinol>	-do-	α-glucosidase inhibitor
Fucosterol	Sterol	PTP 1B, α-glucosidase, ACE, AGEs, and aldose reductase inhibitor

Li (2016) reported that its methanolic extracts reduced plasma glucose levels in experimental rats, and its phlorotannin *viz.* phlorofucofuroeckol-A had AGEs inhibitory activity. Its water extracts showed α-glucosidase inhibition against α-glucosidase (Saccharomyces), rat intestinal maltase, rat intestinal sucrase, rat intestinal isomaltase, and rat intestinal glucoamylase with IC50 values of 0.03, 4.2, 10.1, > 100 and > 100 mg/mL, respectively [23]. Further, its methanolic extracts had similar activities with IC50 values of 0.02, 0.8, 4.1, > 100, and 5.9 mg/mL, respectively. Furthermore, its phloroglucinol derivatives *viz.* dieckol, eckol, dioxinodehydroeckol, 7-Phloroeckol, phlorofucofuroeckol, and phloroglucinol showed aldose reductase inhibitory activity with IC50 values of 42,4, 54.7, 22, 27.5, 125.5, and 72.5, µM respectively. The phlorotannin *viz.* 2-phloroeckol and carotenoids derived from the ethyl acetate extract of this species exhibited antidiabetic activity by significantly reducing the activity of aldose reductase enzymes. Carotenoids registered an IC50 value of 18.94 µM [41]. The antidiabetic compounds of both *Ecklonia stolonifera and Eisenia bicyclis* and their inhibitory activities are shown in Table **3**.

Table 3. Antidiabetic compounds of both Ecklonia stolonifera and Eisenia bicyclis and their inhibitory activities [24] (Ezzat *et al.*2018).

Compound	PTP1B Inhibitory Activity (IC50 , µM)	α-glucosidase Inhibitory Activity (IC50 , µM)
Eckol	2.6	22.8
Phlorofurofucoeckol-A	0.6	1.4
Dieckol	1.2	1.6
7-Phloroeckol	2.1	6.1
Phloroglucinol	55.5	141.2
Dioxinodehydroeckol	30.0	34.6

The fucosterol (24-ethylidene cholesterol) of both *Eisenia bicyclis* and *Ecklonia stolonifera* have been reported to exhibit non-competitive type inhibitory activity against PTP1B [24].

Eisenia bicyclis

Global distribution: It is abundant in the coasts of Japan and Korea, and it is also found in Peru and the North Pacific coast (Fig. **35**).

Ecology: It lives attached to rocks in shallow water areas.

Fig.   (35).   *Ecklonia_bicyclis*   Nordisk   familjebok;   public   domain   ；
https://commons.wikimedia.org/wiki/File:Alger,_Ecklonia_bicyclis,_Nordisk_familjebok.png

Antidiabetics and their mechanisms of action: The fucoxanthin, phlorofucofuroeckol-A, derived from this species displayed a potent inhibitory effect of aldose reductase enzyme with an IC50 value of 6.22 μM. A similar inhibitory effect was also confirmed with its fucosterol in the rat lens [41]. Its eckol, (Fig. **36**) 7-phloroeckol (1-(3′,5′-dihydroxyphenoxy)-7-(2″,4″,6″-tri-hydroxyphenoxy)-2,4,9-trihydroxydibenzo-1,4-Dioxin), and dieckol possessed AGE formation inhibitory activity at 1 mM with percentage values of 96.2, 91.1, and 86.7, respectively [23], dioxinodehydroeckol (Fig. **37**) exerted α-glucosidase inhibitory activity, 7-phloroeckol displayed PTP 1B and α-glucosidase inhibitory activities, fucoxanthin had PTP 1B and aldose reductase inhibitory activity [41], and its dieckol and eckol showed α-amylase inhibitory activity [8].

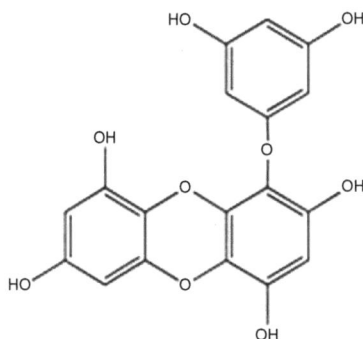

Fig. (36). Eckol.

Eucheuma denticulatum

Global distribution: Philippines, Asia, and the western Pacific.

Ecology: It is known to reside well below the low tide areas to the upper subtidal zones of the coral reef areas.

Antidiabetics and their mechanisms of action: The magnesium (30-90% per 100g dw) present in this algal species possessed hypoglycaemic activity (Fig. **38**) [6].

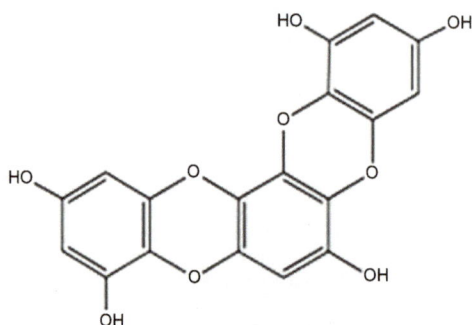

Fig. (37). Dioxinodehydroeckol.

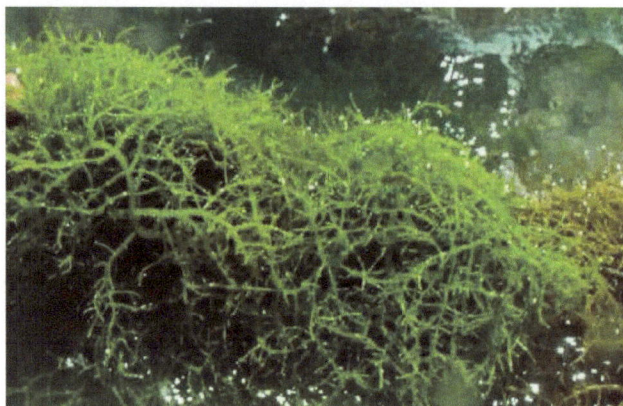

Fig. (38). *Eucheuma_denticulatum.* StinaTano; Creative Commons Attribution-Share Alike 3.0 Unported license.; https://commons.wikimedia.org/wiki/File:Eucheuma_denticulatum_in_an_off-bottom_cultivation,_Bweleo,_Zanzibar.JPG

Fucus distichus

Global distribution: Northern Hemisphere.

Ecology: It is a sessile species and is seen attached to bedrock on exposed rocky shores (Fig. **39**).

Antidiabetics and their mechanisms of action: The phlorotannin (Fig. **40**) derived from this species has shown α-amylase and α-glucosidase inhibitory activities with IC50 values of 13.9 and 0.89 μg/mL, respectively [23].

Fig. (39). *Fucus_distichus*. Ryan Hodnett; Creative Commons Attribution-Share Alike 4.0 International license.;   https://commons.wikimedia.org/wiki/File:Rockweed_%28Fucus_distichus%29_-_Mobile,_Newfoundland_2019-08-09.jpg

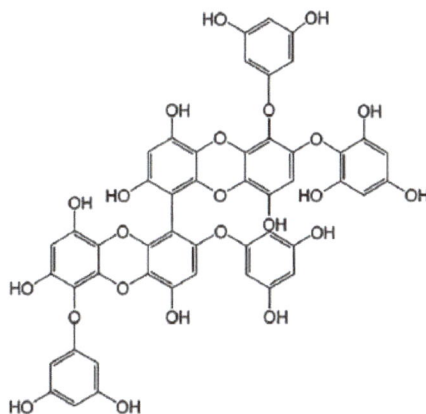

Fig. (40). Phlorotannin (2,700-phloroglucinol-6,60 -biecckol).

Fucus vesiculosus

Global distribution: North Atlantic: From southern Greenland and the White Sea to North Africa.

Ecology: It grows in the littoral zone and the sublittoral zone.

Antidiabetics and their mechanisms of action: The semi-purified phlorotannin fraction of this species showed α-amylase and α-glucosidase inhibitory activities with IC50 values of 2.8 and 0.82 µg/mL, respectively [18]. The cold water and ethanol extracts of this species possessed α-glucosidase inhibitory activities with IC50 values of 0.32 and 0.49 µg/mL, respectively (Fig. **41**) [23].

Fig. (41). *Fucus_vesiculosus.* Anne Burgess; Creative Commons Attribution-Share Alike 2.0 Generic license. https://commons.wikimedia.org/wiki/File:Bladder_Wrack_%28Fucus_vesiculosus%29_-_geograph.org.uk-
-_224125.jpg

Himanthalia elongata

Global distribution: European Atlantic coasts.

Ecology: It grows in rock pools and on rocks of the fairly wave-exposed shores.

Antidiabetics and their mechanisms of action: The crude polysaccharides and fucan derived from this species showed a reduction of post-prandial blood glucose level in alloxan-induced diabetic rabbits (Fig. **42**) [23].

Fig. (42). *Himanthalia elongata.* Zeewieren; Creative Commons Attribution-Share Alike 3.0 Unported, 2.5 Generic, 2.0 Generic and 1.0 Generic license. https://commons.wikimedia.org/wiki/File:Zeespaghetti.jpg

Hormophysa cuneiformis

Global distribution: It is a common species of the Red Sea.

Ecology: It is found in upper subtidal areas.

Antidiabetics and their mechanisms of action: The methanol extracts of this algal species exhibited α-glucosidase inhibitory activity with an IC50 value of 676.9 μg/mL and a percentage value of 53% [18].

Ishige foliacea

Global distribution: It is found distributed along the coasts of the North Pacific Ocean, as well as in Japan, Korea, and China.

Ecology: It is abundantly found on rocks in the upper and middle intertidal zones of open coasts.

Antidiabetics and their mechanisms of action: A phenolic compound, octaphlorethol A (Fig. 43), isolated from this algal species showed antidiabetic effects through the enhancement of glucose uptake by GLUT4 *via* PI3-K/Akt and AMPK signaling pathways [45]. Its octaphlorethol A possessed much higher α-glucosidase inhibitory activity [6].

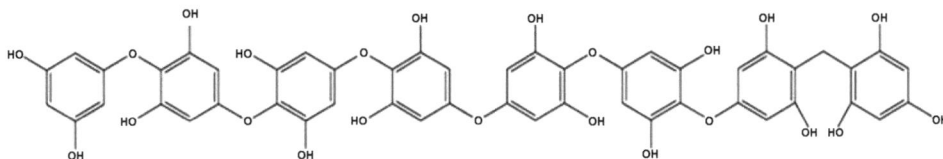

Fig. (43). Octaphlorethol A.

Ishige okamurae

Global distribution: Coasts of the North Pacific Ocean.

Ecology: It is abundantly seen on rocks in the upper and middle intertidal zones of open coasts.

Antidiabetics and their mechanisms of action: Its diphlorethohydroxycarmalol (DPHC) (Fig. 44), a type of phlorotannin, has been reported to possess both α-amylase and α-glucosidase inhibitory activities with IC50 values of 0.53 and 0.16 mM, respectively [41]. The DPHC repressed high glucose-induced angiogenesis *in vitro* in human vascular endothelial cells by suppressing VEGFR-2(vascular endothelial growth factor receptor 2). Based on this finding, it is believed that the DPHC may serve as a promising therapeutic agent for angiogenesis induced by diabetes. Further, the compound ishophloroglucin A (Fig. 45) isolated from this species has shown the highest α-glucosidase inhibitory effect [23].

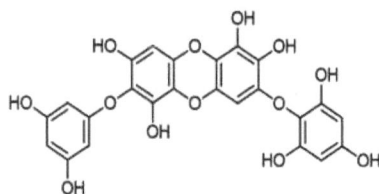

Fig. (44). Diphlorethohydroxycarmalol.

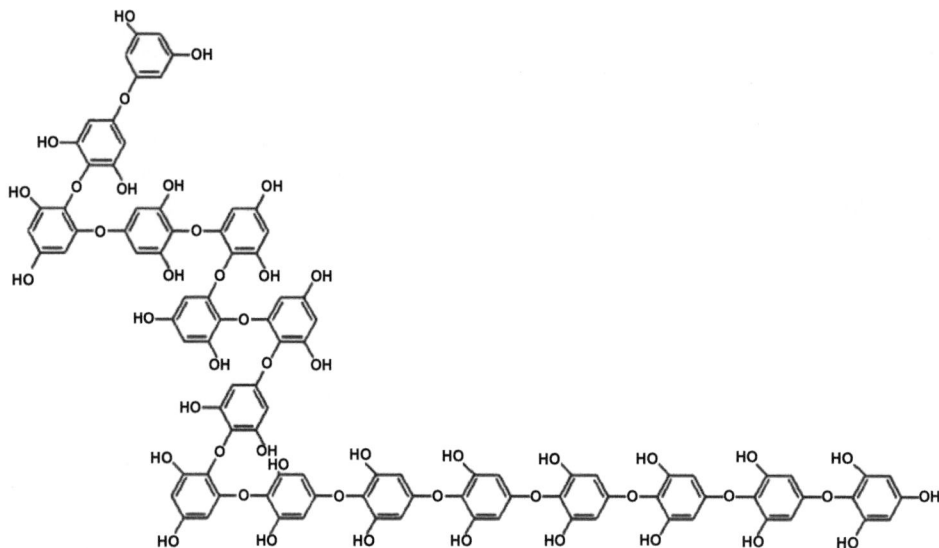

Fig. (45). Ishophloroglucin A.

Laminaria digitata

Global distribution: Coasts of Britain and Ireland, and North Sea coasts of Scandinavia.

Ecology: It is found in moderately exposed sublittoral areas (Fig. **46**).

Antidiabetics and their mechanisms of action: The algal polysaccharide *viz.* high viscous alginates (Fig. **47**) of this species reduced the blood glucose in experimental pigs. These alginates were found to lead to decreased glucose absorption (up to 50% over a period of 8 h) by the reduction of blood glucose [6].

Myagropsis myagroides

Global distribution: It is a common species on the coast of East Asia.

Ecology: It grows in the subtidal zone.

Antidiabetics and their mechanisms of action: The plastoquinone, sargahydroquinoic acid (Fig. **48**), derived from this species has shown both PTP 1B and ACE inhibitory activities [41].

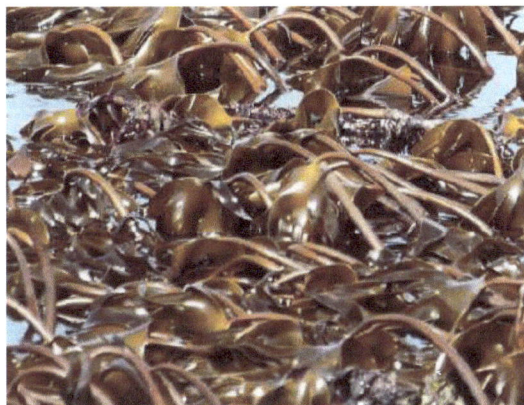

Fig. (46). *Laminaria_digitata.* Hans Hillewaert; Creative Commons Attribution-Share Alike 4.0 International license.; https://commons.wikimedia.org/wiki/File:Laminaria_digitata_%28at_low_tide%29.jpg

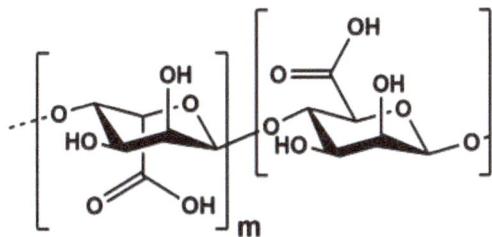

Fig. (47). Alginates.

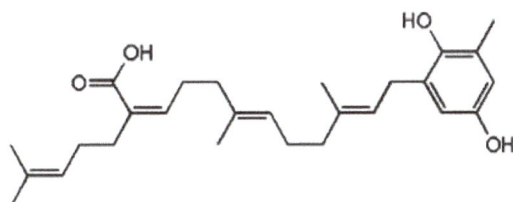

Fig. (48). Sargahydroquinoic acid.

Padina antillarum

Global distribution: Western Central Atlantic and Indo-Pacific.

Ecology: It is found in the coastal regions within rocky intertidal or submerged reef-like habitats.

Antidiabetics and their mechanisms of action: The MeOH extracts (80%) of this species demonstrated antidiabetic activity by inhibiting the alpha-glucosidase with an IC50 value of 13.1 ug/mL [46].

Padina boergesenii

Global distribution: Tropical and subtropical western Atlantic Ocean, Mediterranean Sea, and Pacific Ocean.

Ecology: Its habitat ranges from intertidal to subtidal zones.

Antidiabetics and their mechanisms of action: The aqueous extracts of this species displayed antidiabetic activity through the reduction of elevated blood glucose levels (Fig. **49**) [18].

Fig. (49). *Padina_boergesenii.* James St. John; Creative Commons Attribution 2.0 Generic license.; https://commons.wikimedia.org/wiki/File:Padina_boergesenii_(leafy_rolled-blade_algae_Bahamas.jpg

Padina gymnospora

Global distribution: It has a worldwide distribution, especially in the tropics and on the East and West coasts of Australia.

Ecology: Its habitat ranges from intertidal to subtidal zones.

Antidiabetics and their mechanisms of action: The hexanic extract of this species exhibited α-glucosidase inhibitory activity with an IC50 value of 36.9 μg/mL [47].

Padina sulcata

Antidiabetics and their mechanisms of action: The extract of this species has shown antidiabetic effect at a concentration of 10 mg/ml by significantly inhibiting the DPP-4 enzyme with a percentage value of 83.1 [42].

Pelvetiopsis arborescens

Global distribution: It is found widely distributed in warm-temperate and tropical regions.

Ecology: It resides in the intertidal and subtidal areas up to 2 m deep.

Antidiabetics and their mechanisms of action: The methanolic extracts of this species have shown α-glucosidase and α-amylase inhibitory activities with IC50 values of 0.26 and 0.23 mg/mL, respectively (Fig. **50**) [18].

Fig. (50). *Pelvetiopsis* sp.; Peter D. Tillman; Creative Commons Attribution 3.0 Unported license.; https://commons.wikimedia.org/wiki/File:Dwarf_rockweed,_north_Moonstone.jpg

Polycladia myrica (= Cystoseira myrica)

Global distribution: Mediterranean and Northeast Atlantic.

Ecology: It is a coastal water species and is commonly seen in the intertidal and subtidal rocky reefs.

Antidiabetics and their mechanisms of action:

The 80% MeOH extracts of this species exhibited antidiabetic activity with the inhibition of alpha-glucosidase with an IC50 value of 12.7 ug/mL [46]. Diabetic rats treated with hot aqueous extract of this algal species for 30 days had significantly lowered blood glucose levels *i.e.* from 465 mg/dl to 255 mg/dl [48].

Saccharina angustata (= Laminaria angustata)

Global distribution: Northern Atlantic; Japan, China, Taiwan, Korea, and Philippines.

Ecology: It colonizes the rocky substrates of the coastal areas.

Antidiabetics and their mechanisms of action: The raw extracts of this species reduced the plasma glucose levels in experimental rats [8].

Saccharina japonica (= Laminaria japonica)

Global distribution: North Western Pacific: Japan, Hokkaido, Korea, China; Mediterranean: France.

Ecology: It grows in subtidal areas from the upper infralittoral level.

Antidiabetics and their mechanisms of action: The butyl-isobutyl-phthalate (Fig. **51**) derived from this algal species has shown α-glucosidase inhibitory activity with an IC50 value of 38.00 μM. Further, this compound displayed a hypoglycaemic effect in streptozotocin-induced diabetic mice. Furthermore, its porphyrin derivative pheophorbide A (Fig. **52**) displayed the inhibitory activity of AGE formation and aldose reductase with IC50 values of 49.43 and 12.31 μM, respectively, and its pheophytin A (Fig. **53**) had similar activities with IC50 values of 228.71 and > 100 μM, respectively [23 - 41].

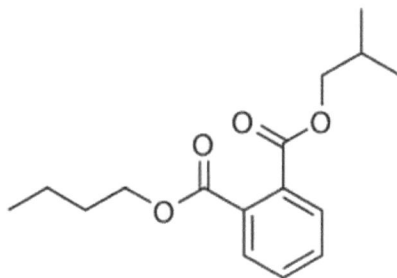

Fig. (51). Butyl-isobutyl-phthalate.

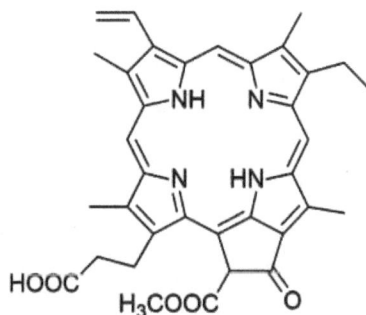

Fig. (52). Pheophorbide-A.

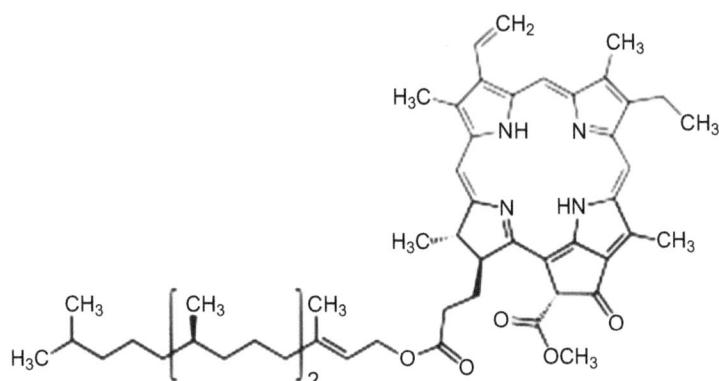

Fig. (53). Pheophytin-A.

Sargassum angustifolium

Global distribution: Temperate and tropical oceans of the world; it is a common species on the southwestern coast of Persian Gulf.

Ecology: It is known to occur in shallow water areas with coral reefs.

Antidiabetics and their mechanisms of action: Diabetic mice fed with free and encapsulated fucoxanthin (400 mg/kg) showed decreased fasting blood glucose and increased plasma insulin level [49].

Sargassum aquifolium (= Sargassum binderi)

Global distribution: Indian Ocean, Indonesia, Kenya, Madagascar, Mauritius, Samoa, Singapore, Taiwan, Tasmania, Fiji, French Polynesia, and Hawaii.

Ecology: It is found in subtidal areas with rocky terrain and reef flats up to a depth of 3 m.

Antidiabetics and their mechanisms of action: The extract of this species showed antidiabetic effects at a concentration of 10 mg/ml by potently inhibiting the DPP-4 enzyme with a percentage value of 81.8 [42]. The ethanolic precipitates of this algal species displayed DPP-4 inhibitory activity with an IC50 value of 2.2 mg/mL [23].

Sargassum boveanum

Global distribution: Western Indian Ocean.

Ecology: It generally inhabits shallow water and coral reefs.

Antidiabetics and their mechanisms of action: The MeOH extracts (80%) of this species showed antidiabetic activity by inhibiting alpha-glucosidase enzyme with an IC50 value of 19.7 ug/mL [46].

Sargassum buxifolium

Global distribution: Tropical and subtropical western Atlantic.

Ecology: It is commonly seen in shallow coral reef areas.

Antidiabetics and their mechanisms of action: The hexanic extracts of this species show α-glucosidase inhibitory activity with an IC50 value of 36.9 μg/ml [47].

Sargassum confusum

Global distribution: Tropical and subtropical coastal areas.

Ecology: It is known to grow on rocks in areas from the intertidal to the subtidal zone.

Antidiabetics and their mechanisms of action: The administration of oligosaccharides of this species in HFSD-fed hamsters led to decreased fasting blood glucose. This antidiabetic effect was found to be due to the regulation of c-Jun N-terminal kinase and insulin receptor substrate 1/phosphatidylinositol 3-kinase pathways. Further, the extracts of this species have also shown anti-diabetic activity in HepG2 cells in hamsters [6].

Sargassum fusiforme

Global distribution: Yellow Sea and Bohai Sea of China, as well as Japan, South Korea, and North Korea.

Ecology: It generally inhabits shallow coral reef areas.

Antidiabetics and their mechanisms of action: The polysaccharides of this algal species have shown extremely potent alpha-glucosidase inhibitory activity [50]. The polysaccharides of this species administered orally mitigated hyperglycemia and hyperinsulinemia in experimental animals. The anti-diabetic effects of its polysaccharides are believed to be due to the rapid absorption and use of blood glucose in the muscle and liver [51].

Sargassum glaucescens

Global distribution: Tropical and subtropical western Pacific, including Japan and SE Asian countries.

Ecology: It is commonly seen on rocky coasts.

Antidiabetics and their mechanisms of action: A total of six sterols, *viz.* β-sitosterol, stigmasterol, 24(S)-hydroxy-24-vinylcholesterol, (Fig. **54**) fucosterol, 24(R)-hydroxy-24- vinylcholesterol (Fig. **55**), and cholesterol (Fig. **56**) have been derived from this species. Among these sterols, the oral administration of β-sitosterol at 10, 15, and 20 mg/kg in experimental animals was found to decrease the serum glucose, with concomitant increases in serum insulin levels. Further, the extracts of this species have also shown α-amylase inhibitory activities with an IC50 value of 8.9 mg/mL [52].

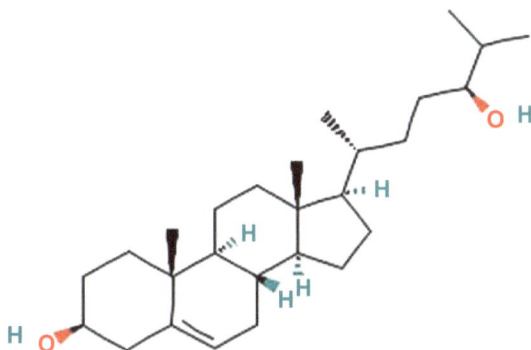

Fig. (54). 24(S)-hydroxy-24-vinylcholesterol.

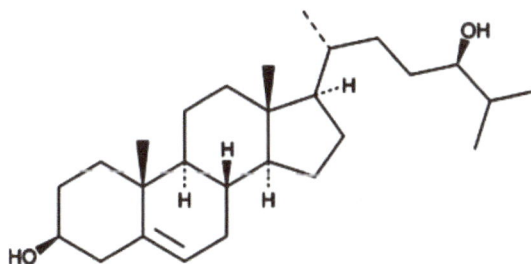

Fig. (55). 24(R)-hydroxy-24- vinylcholesterol.

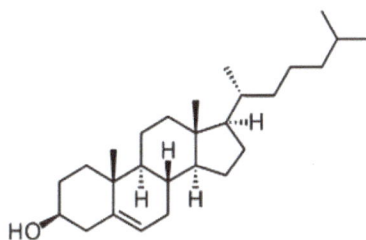

Fig. (56). Cholesterol.

Sargassum hemiphyllum

Antidiabetics and their mechanisms of action: The acetone extracts of this species have shown α-amylase, sucrase, and maltase inhibitory activities with IC50 values of 0.35, 1.89, and 0.09 mg/mL, respectively [23].

Sargassum horneri

Global distribution: It is found widely distributed throughout the coastal waters of Asia.

Ecology: It generally inhabits shallow water and coral reefs.

Antidiabetics and their mechanisms of action: At concentrations of 250, 500, and 1000 μg/mL, the extracts of this algal species showed α-glucosidase inhibitory activity with percentage values of 77.1, 85.7, and 89.6%, respectively (Fig. **57**) [53].

Fig. (57). *Sargassum_horneri.* Totti; Creative Commons Attribution-Share Alike 4.0 International license.; https://commons.wikimedia.org/wiki/File:Sargassum_horneri_Kaikyokan.jpg

Sargassum hystrix

Global distribution: Western Atlantic: Belize and USA.

Ecology: It is known to grow in less illuminated shallow habitats.

Antidiabetics and their mechanisms of action: Orally administrated extracts of this species at 300 mg/kg were found to significantly reduce the preprandial and postprandial glucose levels in STZ-induced diabetic rats [54].

Sargassum pallidum

Global distribution: China, Japan, and other Asian countries.

Ecology: It is seen in rocky marine pools, intertidal zones, coral reefs, and moderately deep coastal waters.

Antidiabetics and their mechanisms of action: The extracts of this species have been reported to alleviate hyperglycemia and insulin resistance. Further, these extracts were found to modulate glucose metabolism by regulating the levels of mRNA expression. It is also suggested that this alga may serve as an effective dietary supplement or drug in the management of T2DM [55].

Sargassum patens

Global distribution: Atlantic, Pacific, and Indian Oceans.

Ecology: Its habitat ranges from midlittoral to sublittoral zones.

Antidiabetics and their mechanisms of action: The phloroglucinol derivative, 2-(4-(3,5-dihydroxyphenoxy)-3,5-dihydroxyphenoxy)-benzene-1,3,5-triol (DDBT) (Fig. **58**), derived from this species displayed significant inhibitory activity against carbohydrate-hydrolyzing enzymes [6]. Its DDBT shows α-amylase, maltase, and sucrase inhibitory activities with IC50 values of 3.2, 114, and 25.4 µg/mL, respectively [23]. The DDBT of this species potently suppressed the hydrolysis of amylopectin by human salivary and pancreatic α-amylases. According to these authors, DDBT is a potent inhibitor of carbohydrate-hydrolyzing enzymes, and it may be of great use as a natural nutraceutical for preventing diabetes [56].

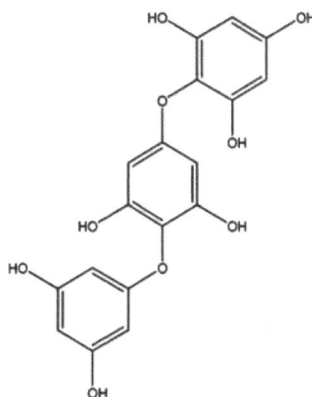

Fig. (58). DDBT.

Sargassum polycystum

Global distribution: Atlantic and Indo-Pacific oceans.

Ecology: It inhabits shallow reef flats and rocky bottoms.

Antidiabetics and their mechanisms of action: Ethyl acetate extract of this species displayed maximum inhibition against α-amylase with an IC50 value of 438.5 µg/ml. On the other hand, its methanol extract exerted potent inhibition against DPP-IV and α-glucosidase with IC50 values of 36.9 and 289.7 µg/ml, respectively [23].

Sargassum ringgoldianum

Global distribution: Pacific coast of Honshu.

Ecology: It generally inhabits shallow water and coral reefs.

Antidiabetics and their mechanisms of action: The methanolic (80%) extracts of this species have been reported to display α-amylase and α-glucosidase inhibitory activities with I50 values of 0.18 and 0.12 mg/mL, respectively. Further, the above extract has also shown delayed absorption of dietary carbohydrates and reduction of post-prandial blood glucose level [23].

Sargassum serratifolium

Global distribution: It is found widely distributed on Korean and Japanese coasts.

Ecology: It occurs in open ocean and unprotected coastal habitats with exposure to wave action and tidal fluctuation.

Antidiabetics and their mechanisms of action: The compounds sargahydroquinoic acid, sargaquinoic acid, and sargachromenol (Figs. **59-61**) extracted from this alga displayed PTP1B inhibitory activity with IC50 values of 5.14, 14.15, and 11.80 µM, respectively [24]. Its sargaquinoic acid served as PPAR (peroxisome proliferator-activated receptor.) agonists and sargachromenol acted both as PPAR agonists and alpha-glucosidase inhibitory agent. PPAR agonists are drugs used for lowering blood sugar [6].

Sargassum swartzii (= Sargassum wightii)

Global distribution: It is found distributed throughout the temperate and tropical oceans of the world.

Ecology: It occurs both in subtidal and intertidal areas.

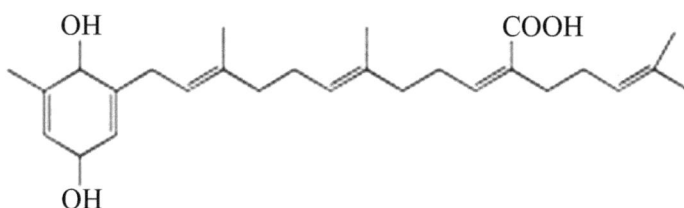

Fig. (59). Sargahydroquinoic acid.

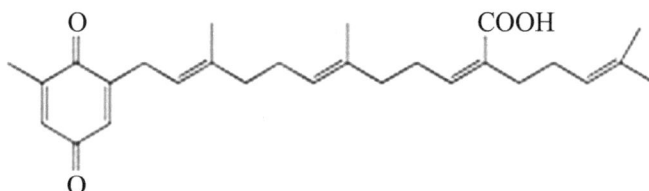

Fig. (60). Sargaquinoic acid.

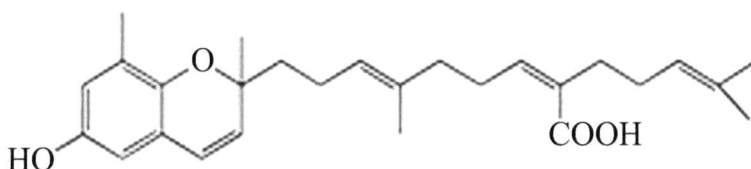

Fig. (61). Sargachromenol.

Antidiabetics and their mechanisms of action: The petroleum ether and ethyl acetate extracts of this algal species have shown potent inhibitory activity against α-glucosidase and α-amylase with IC50 values of 314.8 and 378.3 µg/ml, respectively. Further, its methanol extract displayed the maximum inhibition against dipeptidyl peptidase-IV (DPP-IV) with an IC50 value of 38.27 µg/ml [57].

Silvetia babingtonii (= Pelvetia babingtonii)

Global distribution: It is distributed worldwide, especially in the Pacific Ocean.

Ecology: It is found in the intertidal zone of rocky seashores.

Antidiabetics and their mechanisms of action: The 70% methanolic extract of this species exhibited α-glucosidase and sucrase inhibitory activity in the experimental rats with IC50 values of 2.24 and 2.84 mg/ml. Further, the postprandial elevation in the blood glucose level was found to be significantly suppressed at 15 and 30 min after the administration of sucrose with the extract (Fig. **62**) [58].

Fig. (62). *Silvetia* sp. Patrice78500; Creative Commons Attribution-Share Alike 3.0 Unported license. https://commons.wikimedia.org/wiki/File:Pelvetia_canaliculata_in_Belle-%C3%8Ele-en-Mer.JPG

Silvetia siliquosa (= Pelvetica siliquosa)

Global distribution: It is distributed worldwide, especially in the Pacific Ocean.

Ecology: It is found in the intertidal zone of rocky seashores.

Antidiabetics and their mechanisms of action: The extracts of this algal species enhanced insulin concentration and reduced plasma glucose levels in experimental rats. Oral administration of its fucosterol at 30 mg/kg in streptozotocin-induced diabetic rats caused a potent decrease in the serum glucose concentrations, and at 300 mg/kg, there was a significant reduction in blood glucose level in epinephrine-induced diabetic rats [59].

Spatoglossum schroederi

Global distribution: It is found distributed in the Pacific, Atlantic, and Indian oceans.

Ecology: It is found in all habitats, such as rocky fringing reefs, sandy fringing reefs, patch reefs, estuaries, and coasts.

Antidiabetics and their mechanisms of action: The acetone crude extract of this species showed α-amylase inhibitory activity with an ED50 value of 0.58 mg/mL (Fig. **63**) [23].

Fig. (63). *Spatoglossum* sp.; Wendy Nelson; Creative Commons Attribution 4.0 International license.; https://commons.wikimedia.org/wiki/File:Spatoglossum_chapmanii_Lindauer_(AM_AK331441).jpg

Taonia atomaria

Global distribution: Western Ireland and SW Britain; Mediterranean.

Ecology: It resides in the sunny, sandy pools in the lower intertidal areas.

Antidiabetics and their mechanisms of action: The ethanol crude extract of this species showed α-amylase inhibitory activity with a percentage value of 66.3 (Fig. **64**) [18].

Fig. (64). *Taonia_atomaria-*. Pierre-Louis Crouan (1798-1871) & Hippolyte-Marie Crouan (1802-1871); public domain; https://commons.wikimedia.org/wiki/File:Taonia_atomaria_Crouan.jpg

Turbinaria conoides

Global distribution: Indian Ocean.

Ecology: It is mostly seen on sandy coralline bottoms on reef portions.

Antidiabetics and their mechanisms of action: Its extract showed a potent inhibitory effect on the DPP-4 enzyme at a concentration of 10 mg/ml with a percentage value of 76.2. Further, this species was found to secrete a glucagon-like peptide-1 (GLP-1), which is known to prevent hyperglycaemic conditions [42]. The ethanolic precipitates and aqueous extracts of this species showed DPP-4 inhibitory effects (IC50 = 3.594 mg/mL) and Gastric Inhibitory Polypeptide (GIP) secretory activity, respectively (Fig. **65**) [23].

Fig. (65). *Turbinaria* sp. ; Anne Hoggett; Creative Commons Attribution 3.0 Unported license.; https://commons.wikimedia.org/wiki/File:Turbinaria_ornata_%28Cnidaria%29.jpg

Turbinaria decurrens

Global distribution: Indo-West Pacific Ocean.

Ecology: This sessile species is found attached to rocks in wave-exposed habitats from lower intertidal to upper subtidal areas.

Antidiabetics and their mechanisms of action: The acetone extracts of this species exhibited α-glucosidase and α-amylase inhibitory activities with IC50 values of 2.8 and 4.4 ug/mL, respectively, and percentage values of 96.1 and 97.4, respectively. Further, its methanol crude extract was found to show α-glucosidase inhibitory activity with an IC50 value of 11 μg/mL [18].

Turbinaria ornata

Global distribution: Throughout the Pacific and Indian Ocean.

Ecology: This intertidal, sessile species is largely found attached to the crevices of rocks as well as in the crevices of coral heads at 20-30 m deep.

Antidiabetics and their mechanisms of action: The methanol extract of this species containing fucoids and sulfated polysaccharides displayed a potent inhibitory effect on the DPP-4 enzyme by 55.4% at 80 µg/ml [42].

Undaria pinnatifida

Global distribution: Northwest Pacific: Vladivostok, Russia; Japan, Korea, China, and Hong Kong.

Ecology: It occupies a wide range of shores from low tide levels up to 15 m.

Antidiabetics and their mechanisms of action: The fucoxanthin derived from this species has been reported to serve as a competitive inhibitor of the aldose reductase enzyme, which is known to be present in all target tissues, causing diabetic complications [42]. Consumption of this seaweed at 50 g day/day reduced blood glucose levels considerably in diabetic patients [6] (Fig. **66**).

Fig. (66). *Undaria_pinnatifida.* Wilcox, Mike; Creative Commons Attribution 4.0 International license.; https://commons.wikimedia.org/wiki/File:Undaria_pinnatifida_%28Harv.%29_Suringar_%28AM_AK35379 6-1%29.jpg

Red Algae

Acanthophora spicifera

Global distribution: Tropical and subtropical regions.

Ecology: It occurs in the shallow reef flats and both tidal and subtidal zones at a depth range of 1-22m.

Antidiabetics and their mechanisms of action: Oral administration of flavonoid-rich extracts of this species for 21 days at 100 mg/kg body weight showed maximum antihyperglycemic activity with a considerable reduction in blood glucose levels [60]. The extracts of this species showed an inhibitory effect on α-amylase and α-glucosidase at a concentration of 100 µg/ml, and the percentage values of inhibition were 54.7 and 46. 9, respectively [42].

Actinotrichia fragilis

Global distribution: Indonesia: Seribu Island, Badi Island, Lombok, and Malang.

Ecology: It forms hemispherical clumps on rocks in subtidal zones with moderate to strong water movement.

Antidiabetics and their mechanisms of action: The crude extracts of this species have shown α-glucosidase inhibitory activity with an IC50 value of 1.3 µg/mL (Sabarianandh *et al.*, 2020) (Fig. **67**).

Fig. (67). *Actinotrichia_fragilis* -Philippe Bourjon; Creative Commons Attribution-Share Alike 3.0 Unported license; https://commons.wikimedia.org/wiki/File:Actinotrichia_fragilis.jpg

Champia parvula

Global distribution: Southest England and western Ireland; Channel Islands.

Ecology: It is seen in subtidal areas and lower intertidal pools.

Antidiabetics and their mechanisms of action: The crude extracts of this species have shown α-amylase inhibitory activity with an IC50 value of 173 µg/ml (Fig. **68**) [61].

Fig. (68). *Champia_parvula.* Voctir, Creative Commons Attribution-Share Alike 4.0 International license. https://commons.wikimedia.org/wiki/File:Champia_parvula.jpg

Gracillaria corticate

Global distribution: Asia, South America, Africa, and Oceania.

Ecology: Its discoid holdfast is found attached to rocky substrates.

Antidiabetics and their mechanisms of action: The extracts of this species showed an inhibitory effect on α-amylase and α-glucosidase at a concentration of 100 µg/ml, with percentage values of 84.7 and 73.5, respectively (Fig. **69**) [42].

Gracillaria edulis

Global distribution: Tropical and subtropical regions: Caribbean and Gulf of Mexico.

Ecology: It is seen in protected areas near islands with sandy or rocky bottoms.

Antidiabetics and their mechanisms of action: The ethyl acetate fraction of this species containing 1H-Indole-2-carboxylic acid,6-(4-ethoxy phenyl)-3-methyl 4-oxo- 4,5,6,7-tetrahydro-isopropyl ester (Fig. **70**) displayed potent α-amylase and α-glucosidase inhibitory activities with IC50 values of 279.5 and 87.9 µg/ml, respectively, and percentage values of 87.9 and 79.6, respectively [42].

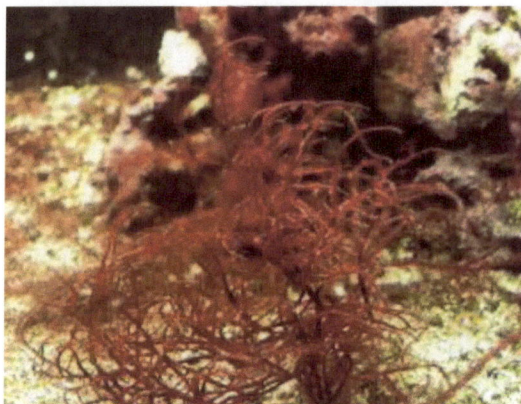

Fig. (69). *Gracilaria* sp.- Emoody26 at English Wikipedia; Creative Commons Attribution 3.0 Unported license. https://commons.wikimedia.org/wiki/File:Gracilaria2.JPG

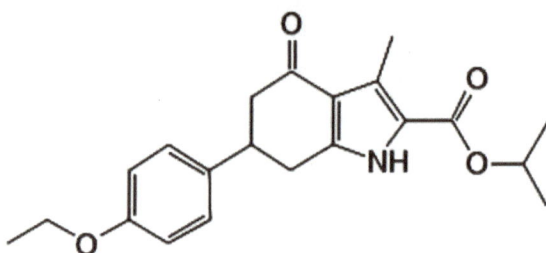

Fig. (70). 1H-Indole-2-carboxylic acid,6-(4-ethoxy phenyl)-3-methyl-4-oxo- 4,5,6,7-tetrahydro-isopropyl ester.

Gracilaria opuntia (name uncertain)

Global distribution: Great Britain, Ireland, Australia, and Chile.

Ecology: It is commonly seen in sheltered areas like bays, estuaries, mangroves, and reefs.

Antidiabetics and their mechanisms of action: The sulfated polysaccharides of this species have been reported to possess alpha-amylase, alpha-glucosidase, and DPP-4 inhibitory activities [62]. (Carpena *et al.*, 2023). The extracts of this species increased the level of insulin at 125 mg/kg BW, which is due to the increased insulin secretion from pancreatic β-cells [61]. Further, the sulfated polygalactans derived from this species have shown inhibitory effects on the DPP-4 enzyme with an IC50 value of 0.09 mg/ml [42].

Grateloupia elliptica

Global distribution: It is found mainly distributed across Japan and Korea.

Ecology: It is commonly seen in upper subtidal areas (Fig. **71**).

Fig. (71). *Grateloupia_elliptica.* Daderot; Creative Commons CC0 1.0 Universal Public Domain Dedication.; https://commons.wikimedia.org/wiki/File:Grateloupia_elliptica_-
_National_Museum_of_Nature_and_Science,_Tokyo_-_DSC07638.JPG

Antidiabetics and their mechanisms of action: The compound 2,4,6-tribromophenol (Fig. **72**) produced by this species has shown α-glucosidase inhibitory activity against Bacillus stearothermophilus, Saccharomyces cerevisiae, rat intestinal maltase, and rat intestinal sucrase with IC50 values of 130.3 μM, 60.3 μM, 5.0 mM, and 4.2mM, respectively. On the other hand, its 2,4-dibromophenol (Fig. **73**) exhibited similar activities against the above with IC50 values of 230.3 μM, 110.4 μM, 4.8 mM, and 3.6 mM respectively [23 - 6].

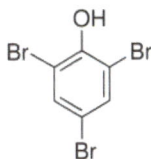

Fig. (72). 2,4,6-tribromophenol.

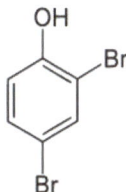

Fig. (73). 2,4-dibromophenol.

Halymenia durvillei

Global distribution: Throughout the temperate and tropical regions of the Atlantic, Indian, Pacific Oceans, and Indo-Pacific regions.

Ecology: It occurs in the lower intertidal zone.

Antidiabetics and their mechanisms of action: The water extract of this species has been reported to possess α-glucosidase inhibitory activity with an IC50 value of 4.34 mg/mL [18 - 63]. The α-glucosidase inhibitory activity of this species was also observed at an IC50 value of 0.3 µg/mL (Fig. **74**) [61].

Fig. (74). *Halymenia* sp. Diaz-Piferrer, M; Creative Commons Attribution 4.0 International license. https://commons.wikimedia.org/wiki/File:Halymenia_floresia_(Clemente)_C.Agardh_(AM_AK339854).jpg

Kappaphycus alvarezii(= Eucheuma cottonii)

Global distribution: Indian and western Pacific Oceans.

Ecology: It is normally seen between the low tide mark and the upper subtidal zone of the coral reef areas (Fig. **75**).

Antidiabetics and their mechanisms of action: The carrageenan (Fig. **76**) derived from this species has shown α-glucosidase inhibitory activity. This antidiabetic activity is largely due to the soluble fiber present in carrageenan, which is believed to reduce the amount of carbohydrates reaching the bloodstream by slowing down the absorption in the small intestine. Further, according to these authors, the magnesium (30-90%) present in this alga showed hypoglycaemic activity [6]. The sulfated polygalactans of this species possess an inhibitory effect on the DPP-4 enzyme with an IC50 value of 0.12 mg/mL. This activity is due to the reaction between the functional groups of sulfated polygalactan with DPP-4 [42].

Fig. (75). *Kappaphycus alvarezii;* Image credit: Prakash Bhuyar *et al*. (Applied for permission).

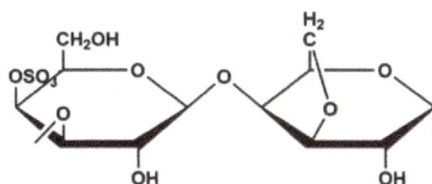

Fig. (76). Carrageenan.

Laurencia dendroidea

Global distribution: Atlantic Ocean.

Ecology: It is seen between the intertidal and subtidal zone at about 3m depth.

Antidiabetics and their mechanisms of action: The ethyl acetate fraction of this algal species exhibited α-glucosidase inhibitory activity with an IC50 value of 8.1 µg/mL [18] (Fig. **77**).

Fig. (77). *Laurencia* sp.; Coughdrop12; Creative Commons Attribution-Share Alike 4.0 International license.; https://commons.wikimedia.org/wiki/File:Laurencia_Aquarium_of_the_Pacific.jpg

Laurencia similis

Global distribution: Temperate and tropical shore areas.

Ecology: It is commonly seen in littoral to sublittoral habitats up to a depth of 65 m.

Antidiabetics and their mechanisms of action: The antidiabetic compounds derived from this species showed PTP1B inhibitory activity [24], and their IC50 values are given in Table **4**.

Table 4. PTP1B inhibitory activity of antidiabetic compounds of *Laurencia similis* [24].

Compound	PTP1B Inhibition (IC50, µM/ µg/mL)
3',5',6',6-Tetrabromo-2,4-dimethyldiphenyl ether (Fig. **78**)	3.0 µM
2',5',6',5,6-Pentabromo-3',4',3,4-tetramethoxybenzo-phenone (Fig. **79**)	2.7 µM
3',5',6'6-Tetrabromo-2,4-dimethyldiphenyl ether (Fig. **80**)	3.0 µg/mL
1,2,5-Tribromo-3-bromoamino-7-bromomethylnaphthalene (Fig. **81**)	102 µg/mL
3',5',6'6-Tetrabromo-2,4-dimethyldiphenyl ether (Fig. **82**)	65.3 µg/mL
2,5,6-Tribromo-3-bromoamino-7-bromomethylnaphthalene (Fig. **83**)	69.8 µg/mL
2',5',6',5,6-Pentabromo-3',4',3,4-tetramethoxybenzo-phenone (Fig. **84**)	2.7 µg/mL

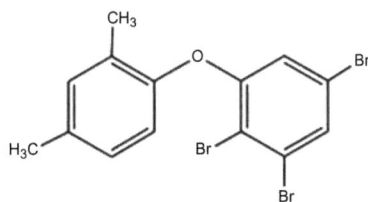

Fig. (78). 3',5',6',6-Tetrabromo-2,4-dimethyldiphenyl ether.

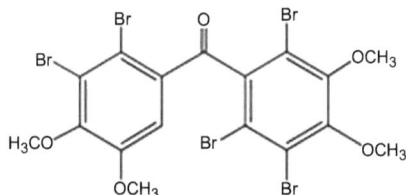

Fig. (79). 2',5',6',5,6-Pentabromo-3',4',3,4-tetramethoxybenzo-phenone.

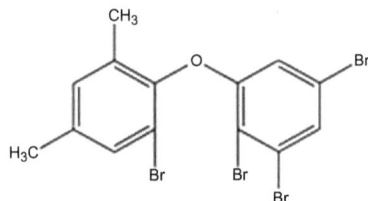

Fig. (80). 3',5',6'6-Tetrabromo-2,4-dimethyldiphenyl ether.

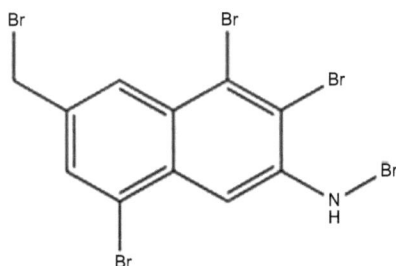

Fig. (81). 1,2,5-Tribromo-3-bromoamino-7-bromomethylnaphthalene.

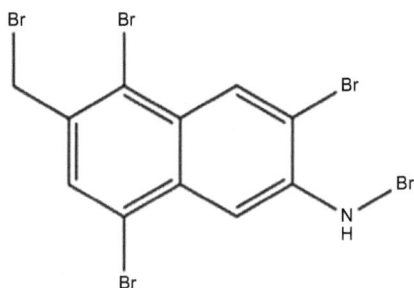

Fig. (82). 2,5,8-Tribromo-3-bromoamino-7-bromomethylnaphthalene.

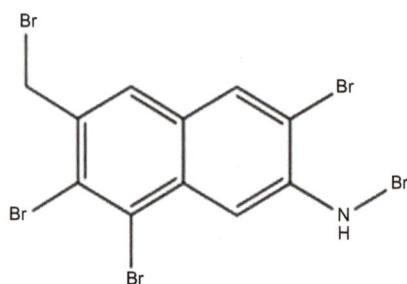

Fig. (83). 2,5,6-Tribromo-3-bromoamino-7-bromomethylnaphthalene.

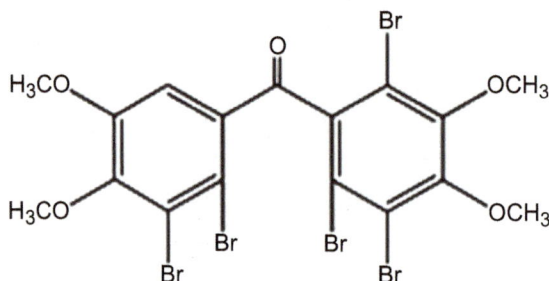

Fig. (84). 2',5',6',5,6-Pentabromo-3',4',3,4-tetramethoxybenzo-phenone.

Odonthalia corymbifera

Global distribution: Not reported.

Ecology: It is a shallow-water species (Fig. **85**).

Antidiabetics and their mechanisms of action: The bromophenols derived from this algal species exhibited antidiabetic activity by inhibiting α-glucosidase [23]. The IC50 values recorded for its various compounds are shown in Table **5**.

Fig. (85). *Odonthalia* sp.; Derek Keats from Johannesburg, South Africa; Creative Commons Attribution 2.0 Generic license.; https://commons.wikimedia.org/wiki/File:Odonthalia_dentata,_Newfoundland_ (7273716074).jpg

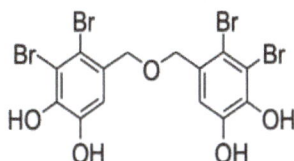

Fig. (86). bis(2,3-dibromo-4,5-dihydroxybenzyl) ether (BDDE).

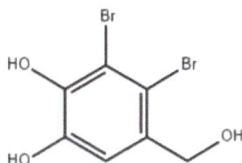

Fig. (87). 2,3-dibromo-4,5-dihydroxybenzyl alcohol.

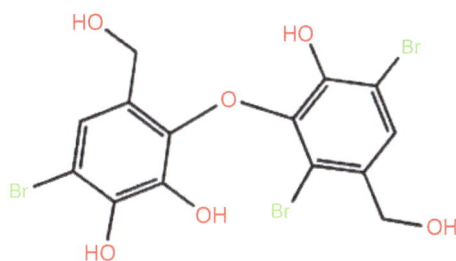

Fig. (88). 4-Bromo-2,3-dihydroxy-6-hydroxymethylphenyl 2,5- dibromo-6-hydroxy-3-hydroxymethylphenyl ether.

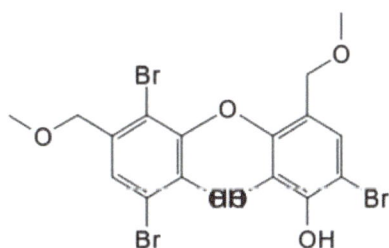

Fig. (89). 4-Bromo-2,3-dihydroxy-6-methoxymethylphenyl 2,5-dibromo-6-hydroxy-3-methoxymethylphenyl ether.

Table 5. α-glucosidase inhibitory activity of bromophenols of *Odonthalia corymbifera* [23].

Compound	α-glucosidase Inhibition (IC50, μM)
bis(2,3-dibromo-4,5-dihydroxybenzyl) ether (Fig. **86**)	0.098
2,5- dibromo-6-hydroxy-3-hydroxymethylphenyl ether (Fig. **87**)	25.0
2,5-dibromo-6-hydroxy-3-methoxymethylphenyl ether, (Fig. **88**)	53.0
2,3-dibromo-4,5-dihydroxybenzyl alcohol (Fig. **89**)	89.0

Palisada perforate (= Laurencia papillosa)

Global distribution: British islands.

Ecology: It is a shallow-water species.

Antidiabetics and their mechanisms of action:

The 80% MeOH extracts of this species exhibited antidiabetic activity with the inhibition of alpha-glucosidase enzyme at an IC50 value of 19.1 ug/mL [46].

Palmaria palmata

Global distribution: Atlantic Europe: from Portugal to the Baltic coasts, as well as the coasts of Iceland and Faroe Islands.

Ecology: This epiphytic species is seen on algae, rock, and mussels in shallow subtidal and intertidal areas up to a depth of about 5 m.

Antidiabetics and their mechanisms of action: The protein hydrolysates of this species have shown dipeptidyl peptidase IV inhibition activity with an IC50 value of 1.65 mg/mL [61]. The aqueous, alkaline, and a mixture of aqueous/alkaline fractions of this species showed DPP-4 inhibitory activity with IC50 values of 2.52, 4.60, and 4.24 mg/ml, respectively (Fig. **90**) [22].

Fig. (90). *Palmaria_palmata.* Pierre-Louis Crouan (1798-1871) & Hippolyte-Marie Crouan (1802-1871); public domain.; https://commons.wikimedia.org/wiki/File:Palmaria_palmata_2_Crouan.jpg

Palmaria sp.

Antidiabetics and their mechanisms of action: The phenolic extracts of this alga showed α-glucosidase and α-amylase inhibitory activities, and an IC50 value of 0.1ug/mL was recorded for its α-amylase inhibitory activity [25].

Polyopes lancifolia (= Gicartina lancifolia)

Global distribution: North Pacific: Japan and adjacent regions.

Ecology: It is found in intertidal and subtidal regions to depths of up to 40 m.

Antidiabetics and their mechanisms of action: A bromophenol compound *viz.* bis(2,3-dibromo-4,5-dihydroxybenzyl)ether (BDDE) derived from this algal species has shown α-glucosidases inhibition against B. stearothermophilus, S. cerevisiae, rat intestinal maltase, and rat intestinal maltase with IC50 values of 0.12 μM, 0.098 μM, 1.20 mM, and 1.00 mM, respectively [23].

Polysiphonia morrowii (uncertain)

Global distribution: North Pacific Ocean.

Ecology: It is known to inhabit the reefs of *Crassostrea gigas* on the Atlantic Patagonian coast (Fig. **91**).

Antidiabetics and their mechanisms of action: The bromophenols, 3-Bromo-4-5-dihydroxy benzyl alcohol (Fig. **92**) and 3-Bromo-4,5-dihydroxybenzyl methyl ether (Fig. **93**), of this algal species showed α-glucosidase inhibitory activity with an IC50 value of 4.8 μg/mL [61].

Fig. (91). *Polysiphonia* sp.; Luis Fernández García ; Creative Commons Attribution-Share Alike 4.0 International license. https://commons.wikimedia.org/wiki/File:Polysiphonia-elongata-19880602a.jpg

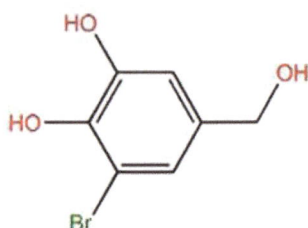

Fig. (92). 3-Bromo-4,5-dihydroxy benzyl alcohol.

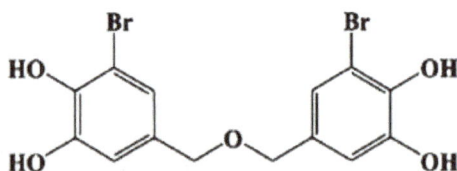

Fig. (93). Bis (3-bromo-4,5-dihydroxybenzyl) ether.

Polysiphonia stricta (= Polysiphonia urceolata)

Global distribution: It is found distributed around the British Isles and West and American Atlantic.

Ecology: This intertidal and subtidal species is commonly seen on rocks and shells; and is epiphytic, especially on Laminaria hyperborea stipes.

Antidiabetics and their mechanisms of action: The bromophenol, 3-Bromo-4(3 bromo-4,5-dihydroxyphenyl methyl)-5-(hydroxymethyl)1,2-benzenediol (Fig. 94), derived from this species showed PTP1B inhibitory activity with an IC50 value of 4.9 µg/mL [61].

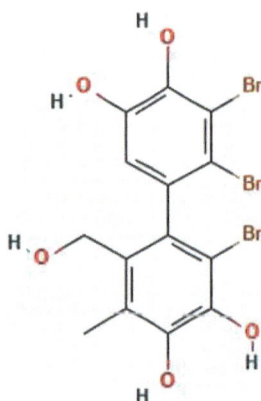

Fig. (94). 3-Bromo-4(3 bromo-4,5-dihydroxyphenyl methyl)-5-(hydroxymethyl)1,2-benzenediol.

Porphyra spp.

Antidiabetics and their mechanisms of action: The bioactive peptides derived from this red alga have been reported to inhibit α-amylase by a noncompetitive binding mode (Fig. **95**) [27].

Fig. (95). *Porphyrs* sp.; Gabriele Kothe-Heinrich; Creative Commons Attribution-Share Alike 3.0 Unported license. https://commons.wikimedia.org/wiki/File:Porphyra_purpurea_Helgoland.JPG

Portieria hornemannii

Global distribution: Tropical and subtropical water bodies of the Pacific; Bermuda, Macaronesia, Mediterranean Sea, and Red Sea.

Ecology: It is a coral reef species.

Antidiabetics and their mechanisms of action: The crude extracts of this species have shown α-amylase inhibitory activity with an IC50 value of 430 μg/mL [61 - 62].

Pyropia yezoensis(= Porphyra yezoensis)

Global distribution: Temperate oceans.

Ecology: It occurs commonly in the intertidal to shallow subtidal areas of coastal waters. Further, it also grows in open coastal areas influenced by cold currents (Fig. **96**).

Antidiabetics and their mechanisms of action: The sulfated poly-galactan porphyrin (Fig. **97**) isolated from this species has been reported to enhance glucose metabolism in diabetic patients by enhancing the level of adiponectin [6 - 63]. The polysaccharides of this species also possessed hypoglycaemic activity, and when these compounds were administered to KKAy mice, they decreased their serum insulin, glucose, and intestinal α-amylase activity, and increased the oral glucose tolerance [64].

Fig. (96). *Pyropia* sp. - Alex Heyman; Creative Commons CC0 1.0 Universal Public Domain Dedication.

Rhodomela confervoides

Global distribution: It is found widespread around the British Isles (Fig. **98**).

Ecology: It is found on rocks and shells and in intertidal pools.

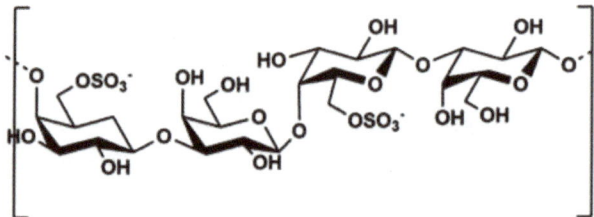

Fig. (97). Porphyrin.

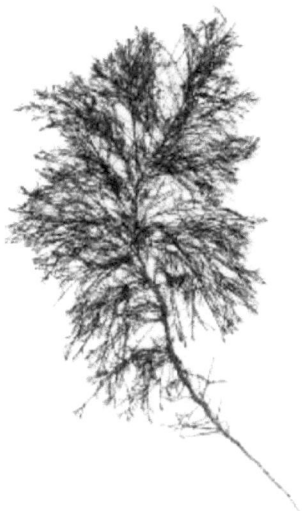

Fig. (98). *Rhodomela_confervoides*. Gabriele Kothe-Heinrich; Creative Commons Attribution-Share Alike 3.0 Unported license.; https://commons.wikimedia.org/wiki/File:Rhodomela_confervoides_Helgoland.JPG

Antidiabetics and their mechanisms of action: The bioactive compound Bis-(2,3 dibromo-4,5-dihydroxybenzyl) ether derived from this algal species displayed α-glucosidase inhibitory activity with an IC50 value of 0.098 µM. Further, its other compounds showed antidiabetic activity through PTP1B inhibition [24 - 25]. The IC50 values recorded for such compounds are shown in Table **6**.

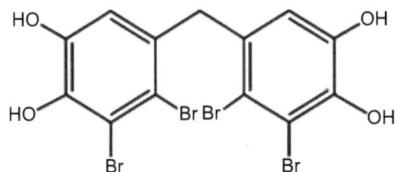

Fig. (99). 2,2′,3,3′-Tetrabromo-4,4′,5,5′-tetra-hydroxydiphenyl methane.

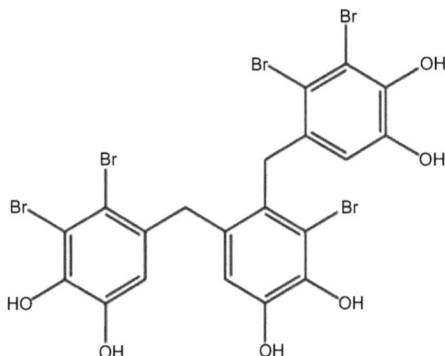

Fig. (100). 3-Bromo-4,5-Bis-(2,3-dibromo-4,5-dihydroxybenzyl)pyrocatechol.

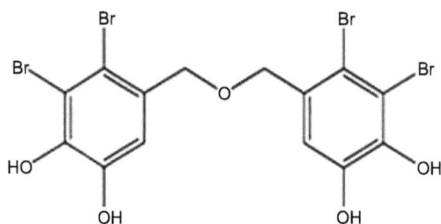

Fig. (101). Bis-(2,3-dibromo-4,5-dihydroxybenzyl) ether.

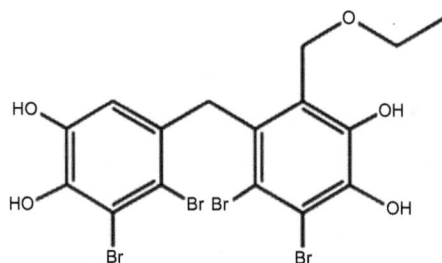

Fig. (102). 2,2′,3,3′-Tetrabromo-3′,4,4′,5-tetrahydroxy-6′-ethyloxymethyldiphenylmethane.

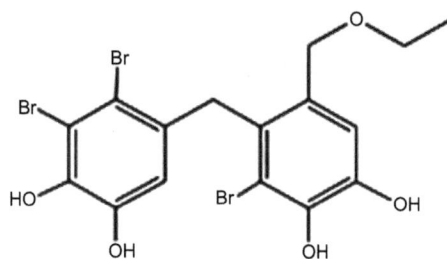

Fig. (103). 3,4-Dibromo-5-(2-bromo-3,4-dihydroxy-6-(ethoxymethyl)benzyl)benzene-1,2-diol.

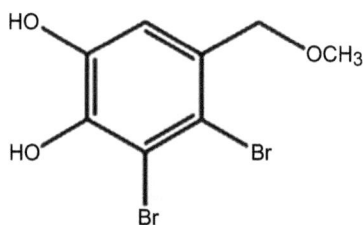

Fig. (104). 3,4-Dibromo-5-(methoxymethyl)benzene-1,2-diol.

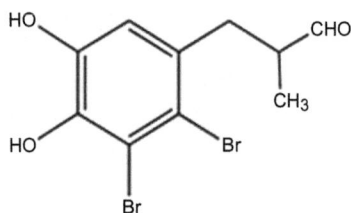

Fig. (105). 3-(2,3-Dibromo-4,5-dihydroxyphenyl)-2-methylpropanal.

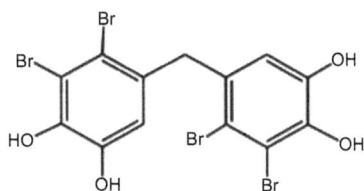

Fig. (106). 3,4-Dibromo-5-(2-bromo-3,4-dihydroxy-6-(isobutoxymethyl)benzyl)benzene-1,2-diol.

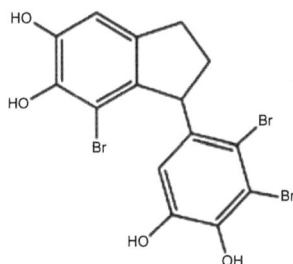

Fig. (107). 7-Bromo-1-(2,3-dibromo-4,5-dihydroxy phenyl)-2,3-dihydro-1H-indene-5,6-diol.

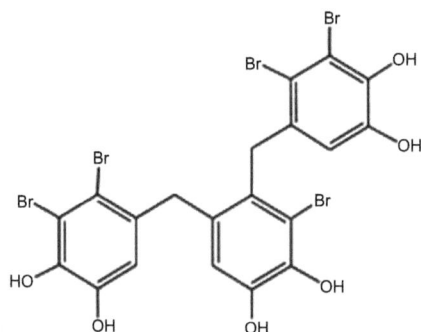

Fig. (108). 5,5'-(3-Bromo-4,5-dihydroxy-1,2-phenylene)-Bis-(methylene))Bis-(3,4- dibromobenzene-1,2-diol).

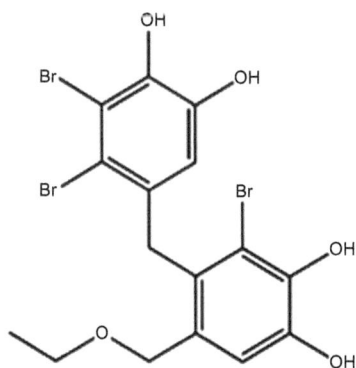

Fig. (109). 3,4-Dibromo-5-(2-bromo-3,4-dihydroxy-6-(ethoxymethyl)benzyl)benzene-1,2-diol.

Table 6. PTP1B inhibitory activity of *Rhodomela confervoides* [24].

Compound	PTP1B Inhibition (IC50, μM)
2,2′,3,3′-Tetrabromo-4,4′,5,5′-tetra-hydroxydiphenyl methane (Fig. **99**)	2.4
3-Bromo-4,5-Bis-(2,3-dibromo-4,5-dihydroxybenzyl)pyrocatechol (Fig. **100**)	1.7
Bis-(2,3-dibromo-4,5-dihydroxybenzyl) ether (Fig. **101**)	1.5
2,2′,3,3′-Tetrabromo-3′,4,4′,5-tetrahydroxy-6′-ethyloxymethyldiphenylmethane (Fig. **102**)	0.8
3,4-Dibromo-5-(2-bromo-3,4-dihydroxy-6-(ethoxymethyl)benzyl)benzene-1,2-diol (Fig. **103**)	0.8
3,4-Dibromo-5-(methoxymethyl)benzene-1,2-diol (Fig. **104**)	3.4
3-(2,3-Dibromo-4,5-dihydroxyphenyl)-2-methylpropanal (Fig. **105**)	4.5
3,4-Dibromo-5-(2-bromo-3,4-dihydroxy-6-(isobutoxymethyl)benzyl)benzene-1,2-diol (Fig. **106**)	2.4
7-Bromo-1-(2,3-dibromo-4,5-dihydroxy phenyl)-2,3-dihydro-1H-indene-5,6-diol (Fig. **107**)	2.8
5,5′-(3-Bromo-4,5-dihydroxy-1,2-phenylene)-Bis-(methylene))Bis-(3,4-dibromobenzene-1,2-diol) (Fig. **108**)	1.7
3,4-Dibromo-5-(2-bromo-3,4-dihydroxy-6-(ethoxymethyl)benzyl)benzene (Fig. **109**)	0.84

Spyridia fusiformis

Global distribution: It is a South Indian species.

Ecology: This coastal species is seen at a depth of 2.5 meters.

Antidiabetics and their mechanisms of action: The crude extract of this algal species showed α-amylase inhibitory activity with an IC50 value of 175 μg/mL [61].

Symphycladia latiuscula

Global distribution: It is found distributed along the coasts of Japan, Korea and China.

Ecology: It is seen in all marine habitats (Fig. **110**).

Antidiabetics and their mechanisms of action: The bromophenol compounds bis (2,3,6-tribromo-4,5 -dihydroxy phenyl) methane, 2,2′,3,5′,6-pentabromo-3′,4,4′,5-tetrahydroxydiphenylmethane, and 2,2′,3,6,6′-pentabromo- 3′,4,4′,5-tetrahydroxydibenzyl ether (Figs. **111-113**) derived from this algal species showed antidiabetic activity by inhibiting aldose reductase [42]. Its bromophenols such as 2,3,6-Tribromo-4,5-dihydroxybenzyl methyl ether, Bis-(2,3,6-tribromo

4,5-dihydroxyphenyl)   methane,   and   1,2-Bis-(2,3,6-tribromo-4,5 dihydroxyphenyl)-ethane showed PTP1B inhibitory activity with IC50 values of 3.9, 4.3, and 2.7 µM, respectively [24].

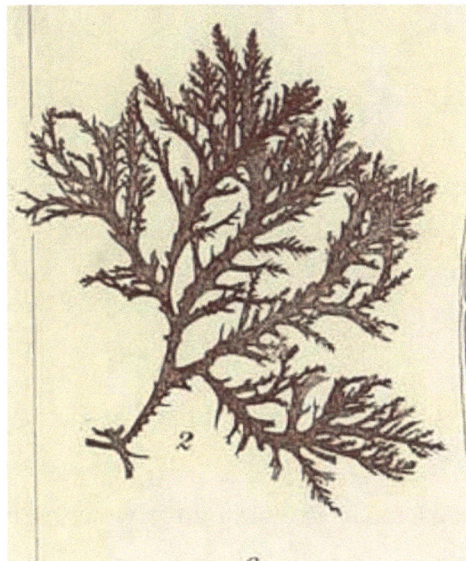

Fig. (110). *Symphycladia* sp.; Okamura, Kintaro;; Creative Commons Attribution 2.0 Generic license.; https://commons.wikimedia.org/wiki/File:Icones_of_Japanese_algae_(Pl._XCVII)_(6049928207).jpg

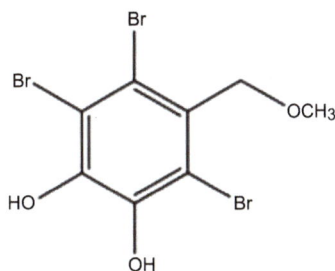

Fig. (111). 2,3,6-Tribromo-4,5-dihydroxybenzyl methyl ether.

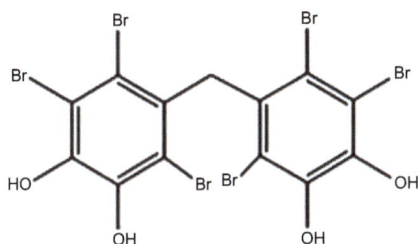

Fig. (112). Bis-(2,3,6-tribromo-4,5-dihydroxyphenyl) methane.

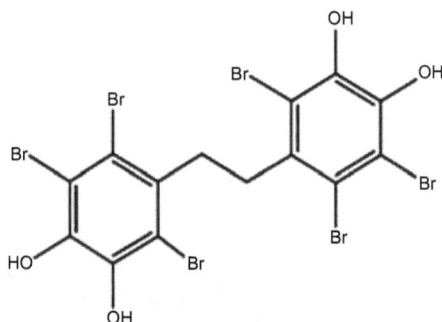

Fig. (113). 1,2-Bis-(2,3,6-tribromo-4,5-dihydroxyphenyl)-ethane.

Seagrasses

Cymodocea nodosa

Global distribution: Mediterranean Sea and parts of the Northeast-Atlantic Ocean, from Portugal to Senegal.

Ecology: It is found restricted to growing underwater and in shallow parts of the sea.

Antidiabetics and their mechanisms of action: The extracts of this species have shown hypoglycemic and hypolipidemic functions. Oral administration of these extracts to diabetic rats inhibited α-amylase and protected the β cells of these rats from death and damage, besides decreasing the blood glucose level by 49% [65] (Kolsi *et al.*, 2017). Its sulfated polysaccharide isolated inhibited α-amylase activity and reduced the blood glucose level by protecting pancreatic β-cells. Further, this compound increases insulin secretion in the blood, leading to improved metabolism and body weight (Fig. **114**) [6].

Halodule uninervis

Global distribution: Asia: Japan, China, Vietnam, and Indonesia.

Ecology: It is a common plant species of the sublittoral zone growing in depths of about 20 m in lagoons, on reefs, and in other types of offshore marine habitats.

Antidiabetics and their mechanisms of action: The methanolic extract of this species has been reported to reduce serum glucose levels in Streptozotocin-induced diabetic rats. The oral administration of this extract at 150 mg/kg reduced glucose levels by 24.8% after 6 h and showed a 52.5% reduction in glucose levels on the 18th day of administration at 150 mg/kg (Fig. **115**) [66].

Fig. (114). *Cymodocea nodosa.* Sylvain Le Bris; Creative Commons Attribution-Share Alike 4.0 International license.; https://commons.wikimedia.org/wiki/File:Cymodocea_nodosa.jpg.

Fig. (115). Halodule uninervis. Paul Asman and Jill Lenoble; Creative Commons Attribution 2.0 Generic license. https://commons.wikimedia.org/wiki/File:Seagrass_Halodule_uninervis_(5777808662).jpg

Halophila beccarii

Global distribution: Indo-Pacific areas.

Ecology: It is abundant in the silty-muddy regions of the intertidal region, as well as sandy and muddy habitats. This shallow coastal waters species also thrives in brackish conditions.

Antidiabetics and their mechanisms of action: The methanol extract of this algal species yielded a 50% inhibition of α-alucosidase and α-amylase at a dose of 100 μg/mL and 270 μg/mL, respectively. Further, this extract was also found to regulate the glucose movement out of the cells [66].

Halophila stipulaceae

Global distribution: It is a native species to the tropical and subtropical waters of the Red Sea, Persian Gulf, and Indian Ocean. It has, however, spread to the Caribbean and Mediterranean seas.

Ecology: It is known to occupy marine sublittoral sediments intertidal to 65 m but is seen mainly at a depth of 30-45 m.

Antidiabetics and their mechanisms of action: At doses of 100 and 200 mg/kg/day, the extracts of this species displayed a 9- and 13-fold increase in serum NO, respectively. This mechanism of action is believed to be due to the improvement of glucose uptake by the tissues *via* the restoration of liver GLUT-2. Further, these extracts were also found to ameliorate oxidative stress status, which is commonly produced by free radicals and dyslipidemia under diabetic situations (Fig. **116**) [66].

Fig. **(116).** *Halophila stipulaceae.* NOAA's National Ocean Service, public domain. https://commons.wikimedia.org/wiki/File:Invasive_seagrass_-_Halophila_stipulacea.jpg

Posidonia oceanica

Global distribution: It is endemic to the Mediterranean Sea.

Ecology: This marine sublittoral vegetation is seen in sandy and silty–sandy sediments.

Antidiabetics and their mechanisms of action: The hydroalcoholic leaf extracts of this algal species displayed antidiabetic activity by inhibiting human serum albumin glycation. Further, no glycation end products were seen by incubating the human serum albumin with glucose in the presence of 0.2 mg of its dry extract for 72 h (Fig. **117**) [66].

Syringodium filiforme

Global distribution: The Caribbean Sea, the Gulf of Mexico, the Bahamas, and Bermuda.

Ecology: It forms meadows in shallow, sandy, or muddy locations. It occurs at a depth of about 20 m.

Antidiabetics and their mechanisms of action: The compounds b-sitosterol and stigmasterol derived from this species were found to possess treatment potential for type 2 DM by increasing GLUT4 translocation and expression (Fig. **118**) [66].

Fig. (117). *Posidonia oceanica.* Frédéric Ducarme; e Creative Commons Attribution-Share Alike 4.0 International license. https://en.wikipedia.org/wiki/File:Posidonia_oceanica_(L).jpg

Fig. (118). *Syringodium filiforme*. James St. John, Creative Commons Attribution 2.0 Generic license. https://commons.wikimedia.org/wiki/File:Syringodium_filiforme_%28manatee_grass%29_%28southeastern_Graham%27s_Harbour,_San_Salvador_Island,_Bahamas%29_1_%2815861656579%29.jpg

Thalassia hemprichii

Global distribution: It is native to the Indian Ocean, the Red Sea, and the western Pacific Ocean.

Ecology: It is mainly found along the coasts.

Antidiabetics and their mechanisms of action: Weight loss is considered to be one of the clinical symptoms of diabetes mellitus, and this is believed to be due to the adipocytes and muscle tissue degeneration and associated conversion of glycogen to glucose. Oral use of the ethanolic extract of this algal species for 15 days resulted in a significant increase in body weight (Fig. **119**) [66].

Fig. (119). *Thalassia hemprichii*. Mudasir Zainuddin, Creative Commons Attribution-Share Alike 4.0 International license. https://commons.wikimedia.org/wiki/File:Thalassia_Hemprichii.jpg

Thalassia testudinum

Global distribution: Cariaco Gulf in northeastern Venezuela.

Ecology: It grows in association with corals between low water tide and a depth of about 10 m.

Antidiabetics and their mechanisms of action: The compounds b-sitosterol and stigmasterol derived from this species possessed treatment potential for type 2 DM by increasing GLUT4 translocation and expression (Fig. **120**) [66].

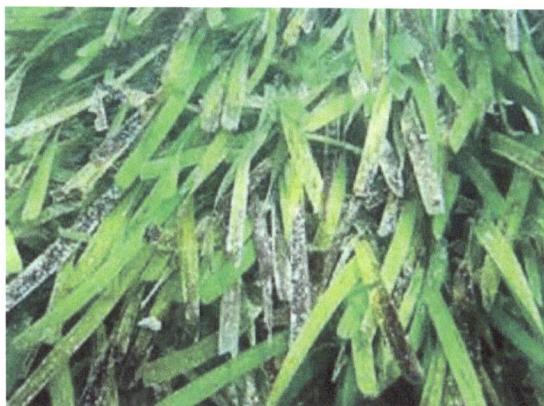

Fig. (**120**). *Thalassia testudinum.* James St. John, Creative Commons Attribution 2.0 Generic license. https://commons.wikimedia.org/wiki/File:Thalassia_testudinum_%28turtle_grass%29_%28South_Pigeon_Cr eek_estuary,_San_Salvador_Island,_Bahamas%29_6_%2815859999657%29.jpg

Zostera subg. *Zostera marina* (= **Zostera marina**)

Global distribution: It is widespread through the Atlantic and Pacific.

Ecology: It is commonly found in mud and sand in protected bays and estuaries.

Antidiabetics and their mechanisms of action: The complex of polar lipids from this species in combination with echinochrome A has been reported to normalize blood glucose of diabetic mice to a controlled level (Fig. **121**) [66, 67].

Mangrove Plants

Acanthus ilicifolius

Global distribution: Asia, Malesia, and islands of the Pacific.

Ecology: It is commonly seen on the coasts of tidal swamps and along the banks of creeks, estuarine islands, and tidal rivers.

Antidiabetics and their mechanisms of action: The flavonoid-containing extracts of this species displayed antidiabetic activity through the regeneration of β-cells of the pancreas [68, 69]. Oral use of 100% ethanol extract of the leaves of this species at 22.4 mg/bw/day significantly reduced blood glucose levels by 69.39% in alloxan-induced experimental animals (Fig. **122**) [70].

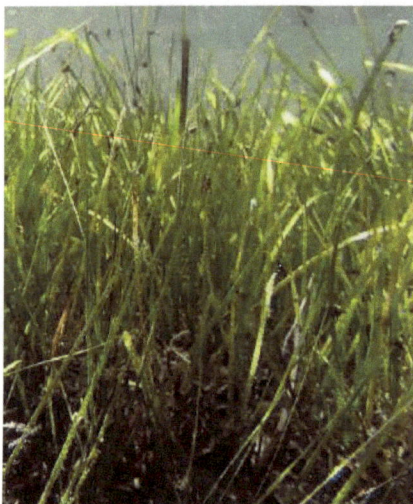

Fig. (121). *Zostera marina.* Sofia Sadogurska; Creative Commons Attribution 4.0 International license. https://commons.wikimedia.org/wiki/File:Seagrass_Zostera_marina_%28Dzharylhach_island%29.jpg

Fig. (122). *Acanthus ilicifolius* . Shagil Kannur ; Creative Commons Attribution-Share Alike 4.0 International license. https://commons.wikimedia.org/wiki/File:Mangroves_at_Muzhappilangad,_Kannur,_Kerala_00_(22).jpg

Acanthus polystachyus

Global distribution: The native range of this species is Ethiopia to NW Tanzania.

Ecology: It is a shrub or tree growing primarily in the seasonally dry tropical biome. It is seen in grassy areas, *e.g.* on roadsides.

Antidiabetics and their mechanisms of action: Oral use of the crude extracts of the root of this plant species daily showed a significant reduction of blood glucose levels in streptozotocin–nicotinamide-induced type 2 diabetic rats [71].

Aegiceras corniculatum

Global distribution: Its distribution range is from India through Southeast Asia to southern China, New Guinea, and Australia.

Ecology: It is seen in coastal and estuarine areas.

Antidiabetics and their mechanisms of action: The presence of alkaloids, flavonoids, saponins, tannins, polyphenols, tannins, and benzofurans in the extracts of this species yielded antidiabetic properties with the utilization of glucose through the mediation of enhanced insulin secretion or by direct stimulation of glucose uptake (Fig. **123**) [69].

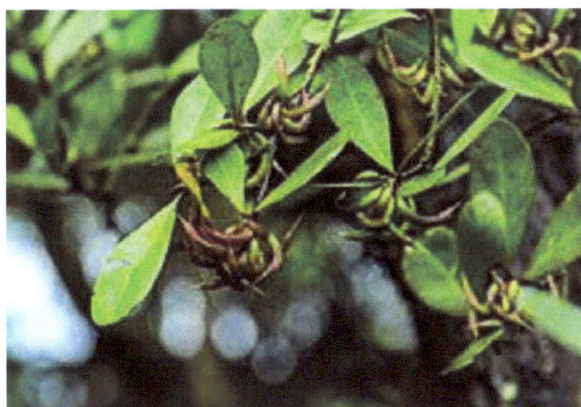

Fig. (123). *Aegiceras corniculatum.* Shagil Kannur; Creative Commons Attribution-Share Alike 4.0 International license.

The oral administration of the extracts of this species at 100 mg/kg showed a moderate reduction in blood glucose (from 382 to 105) and increased activity of liver hexokinase in alloxan diabetic rats [72]. The values of reduction in blood glucose at different extract concentrations are given in Table **7**.

Table 7. Blood glucose levels of alloxan diabetic rats.at different extract concentrations of Aegiceras corniculatum [72].

Experimental Group	Initial Blood Glucose (mg/dl)	Final Blood Glucose (mg/dl)
Diabetic + extract (25 mg/kg)	374.1	120.6
Diabetic + extract (50 mg/kg)	378.8	108.3
Diabetic + extract (100 mg/kg)	381.9	102.9
Diabetic control	362.5	386.6

Aegialitis annulata

Global distribution: It is found in Western Australia and along the coastline of Papua New Guinea.

Ecology: It is a small mangrove growing in exposed sites.

Antidiabetics and their mechanisms of action: The amino acids inorganic salts of this species have been reported to possess antidiabetic properties (Fig. **124**) [69].

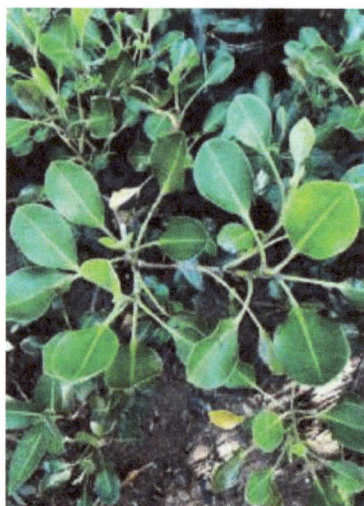

Fig. (124). *Aegialitis annulata.* elawrey; Creative Commons Attribution 4.0 International license. https://commons.wikimedia.org/wiki/File:Aegialitis_annulata_50494002.jpg

Avicennia marina

Global distribution: It is found widely distributed in eastern Africa, the Middle East, Southeast Asia, South Asia, and North to South China. It is also seen in western Australia.

Ecology: It thrives in high-salinity habitats.

Antidiabetics and their mechanisms of action: The saponins present in this species have been reported to display antidiabetic properties through the stimulation of β-cells to release more insulin antiglycation activity [69].

The administration of the extracts of this species to streptozotocin (STZ)-induced diabetic rats was found to result in a significant increase in the serum insulin levels compared to untreated diabetic rats (Fig. **125**) [73].

Fig. (125). *Avicennia marina.* Wie146; Creative Commons Attribution-Share Alike 3.0 Unported, 2.5 Generic, 2.0 Generic and 1.0 Generic license. https://commons.wikimedia.org/wiki/File:Avic_marin_070728_030_mank_rsz.jpg

Avicennia officinalis

Global distribution: Southeast Asia, Malay Archipelago, Australia, and East Asia.

Ecology: It grows on mud flats near the river mouth.

Antidiabetics and their mechanisms of action: The aqueous and petroleum ether (PE) extracts of the leaves of this species show α-amylase and α-glucosidase inhibitory activities, which are due to its phytochemicals including phenols, flavonoids and tannins. The inhibition percentage and IC50 values of the different extracts are shown in Table **8**.

Barringtonia racemosa

Global distribution: It is found distributed from E Africa and Madagascar to Sri Lanka, India, southern China, Taiwan, Myanmar, Thailand, and Andaman and Nicobar Islands (India); it is also found in northern Australia.

Ecology: It is found in rainforest areas and open lowlands. It normally occurs near water, such as along riverbanks and in freshwater swamps. Occasionally, it is seen in the less saline areas of mangrove swamps.

Antidiabetics and their mechanisms of action: The flavonoids, tannins, and saponins present in this species have shown α-glucosidase and α-amylase inhibitory properties [68].

Table 8. α-glucosidase and α-amylase inhibitory activities of *Avicennia officinalis* extracts [68].

Extract Concentration (mg/ml)	% of Alpha-glucosidase Inhibition with PE* Extract	% of Alpha-amylase Inhibition with PE Extract	% of Alpha-glucosidase Inhibition with Water Extract	% of Alpha-amylase Inhibition with Water Extract
0.1	10.0	10.5	15.7	23.2
0.5	13.2	13.0	22.1	27.7
1.0	17.4	17.4	27.4	32.0
IC50	>1mg/ml	>1mg/ml	>1mg/ml	>1mg/ml

* Petroleum ether

Bruguiera cylindrica

Global distribution: Tropical Asia: India, Sri Lanka, Malaysia, Singapore, Philippines, Thailand, Vietnam, Indonesia, New Guinea, and Australia.

Ecology: It is common along estuarine mouths in mangrove forests.

Antidiabetics and their mechanisms of action: The presence of alkaloids, tannins, anthocyanins, flavonoids, phenolic acids, sterols/triterpenoids, *etc.*, in the extracts of this plant species showed antidiabetic effects, which are due to the stimulation of β-cells to release more insulin [69]. The bark extract of this species containing alkaloids, phenolics, flavonoids, and fatty acids exhibited inhibition activity against α-glucosidase and α-amylase with IC50 values similar to that of acarbose, which is a complex oligosaccharide acting as a competitive, reversible inhibitor of intestinal alpha-glucoside hydrolase and pancreatic alpha-amylase. Oral use of 200-400 mg/kg of the extract has been reported to reduce blood glucose levels in experimental rats (Fig. **126**) [74].

Bruguiera gymnorrhiza(= Bruguiera rumphii)

Global distribution: Micronesia, Samoa, and SW Pacific, from the eastern coast of Africa to subtropical Australia.

Ecology: It grows well on dry, well-aerated soil with some freshwater conditions (Fig. **127**).

Fig. (126). *Bruguiera cylindrical*. Shagil Kannur; Creative Commons Attribution-Share Alike 4.0 International license. https://commons.wikimedia.org/wiki/File:Mangroves_at_Muzhappilangad004.jpg

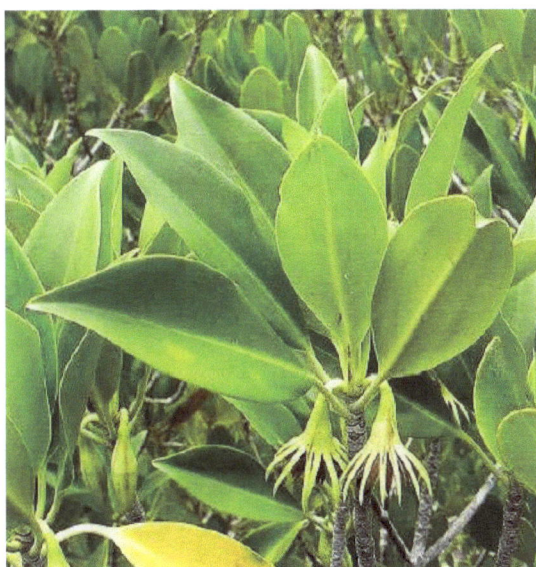

Fig. (127). *Bruguiera gymnorrhiza*. Ton Rulkens from Mozambique; Creative Commons Attribution-Share Alike 2.0 Generic license. https://commons.wikimedia.org/wiki/File:Bruguiera_gymnorrhiza_3_%288349980202%29.jpg

Antidiabetics and their mechanisms of action: The presence of flavonoids, triterpenes, tannins, saponins, polyphenols, and glycosides in the extracts of this species have shown antidiabetic properties [69]. The oral use of the ethanolic extracts of roots at 400 mg/kg body wt reduced blood sugar and anti-hyperglycemic activity in streptozotocin-induced diabetic rats [75], and the values recorded during the different periods of experiment are given in Table **9**.

Table 9. Anti-hyperglycemic activity (Blood glucose level) of ethanolic extracts of Bruguiera gymnorrhiza in streptozotocin-induced diabetic rats [75].

Group	0 Day (mg/ml)	7th Day (mg/ml)	14th Day (mg/ml)	21st Day (mg/ml)	28th Day (mg/ml)
Diabetic control	224.7	214.5	210.0	208.0	201.0
With extract (400mg/kg)	237.0	188.10	129.04	96.0	89.04

Bruguiera parvifolora (not in WoRMS)

Global distribution: South Asia, Indochina, Malesia, and Australia.

Ecology: It is largely seen in the estuarine zones of the high intertidal region (Fig. **128**).

Antidiabetics and their mechanisms of action: The phenolic compounds of this plant species showed antidiabetic properties [69]. Its extracts containing flavonoid compounds *viz.* taxifolin, quercetin, myricetin, kaempferol, and rutin (Figs. **129-131**) displayed significant α-glucosidase inhibitory activities. Among these flavonoids, quercetin exhibited the most promising inhibitory effect with an IC50 value of 3.4 μg/mL (Fig. **128**) [76].

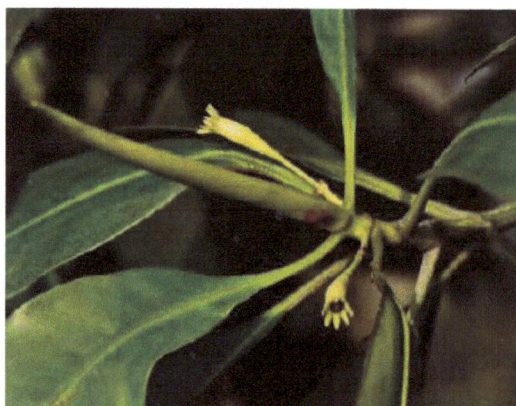

Fig. (128). *Bruguiera parvifolora*. Wibowo Djatmiko (Wie146); Creative Commons Attribution-Share Alike 3.0 Unported license. https://commons.wikimedia.org/wiki/File:Brugu_parvi_111021-18693_Fr_kbu.jpg .

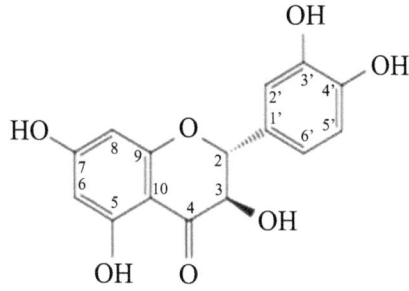

Fig. (129). Taxifolin.

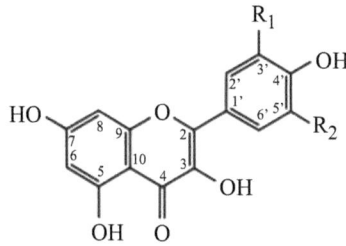

$R_1 = OH, R_2 = H$
$R_1 = R_2 = OH$
$R_1 = R_2 = H$

Fig. (130). Quercetin; Myricetin; Kaempferol.

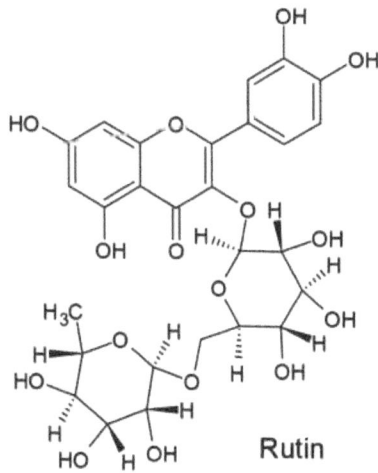

Rutin

Fig. (131). Rutin.

Bruguiera sexangula

Global distribution: Tropical coasts of SE Asia from India to Australia and New Caledonia.

Ecology: It is commonly seen along the outer fringes of tidal forests influenced by freshwater.

Antidiabetics and their mechanisms of action: The alkaloids, steroids, phenolics, tannins, and saponins present in the extracts of this plant species exhibited antidiabetic properties (Fig. **132**) [69].

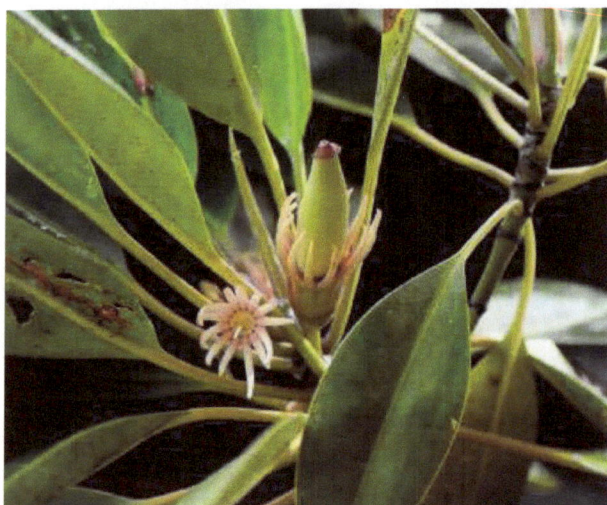

Fig. (132). *Bruguiera sexangular*. Psumuseum; Creative Commons Attribution-Share Alike 3.0 Unported license. https://commons.wikimedia.org/wiki/File:Bruguiera_sexangula.jpg

Ceriops decandra

Global distribution: India, Bangladesh, Burma, Thailand, and Malaysia.

Ecology: It occurs in the tidal creeks and mangrove swamps .

Antidiabetics and their mechanisms of action: The polyphenols, glycosides, flavonoids, tannins, and saponins of this species have shown antidiabetic properties through the stimulation of β-cells to release more insulin and by the increased hexokinase activity [69]. (Sachithanandam *et al*, 2019). The oral administration of 120 mg/kg extract of this plant species was found to potently reduce blood glucose in experimental rats [77]. Its bioactive compound 2-(2 methylphenyl)-1-phenyl-(z)-1-propene exhibited good binding affinities for α-glucosidase, and its 5S*,8S*,9S*,10R*,13S*)-18-hydroxy-16-nor-3-oxod olabr-

4(18)-en-15-oic acid displayed high binding interactions for both α-amylase and α-glucosidase. Further, oral use of the extracts of this plant species at 250 and 500 mg/kg b.w. doses significantly lowered blood glucose levels in experimental mice at 60 and 90-min time points. In the α-glucosidase enzyme inhibitory activity test, its extract registered an IC50 value of 3.37 mg/mL. In the α-amylase enzyme inhibitory activity test, this extract displayed much higher enzyme inhibitory potential with an IC50 value of 2.58 mg/mL (Fig. **133**) [78].

Fig. (133). *Ceriops decandra.* L D Costa, M T H S Budiastuti, J Sutrisno and Sunarto; Creative Commons Attribution-Share Alike 3.0 Unported license. https://commons.wikimedia.org/wiki/File:Ceriops_decandra_flowers.png

Ceriops tagal

Global distribution: Africa, Madagascar, India, Maldives, Seychelles, China, Indo-China, and Australia.

Ecology: It is seen in brackish water areas in tidal zones.

Antidiabetics and their mechanisms of action: The flavonoids, saponins, tannins, and polyphenols present in the extracts of this species yielded antidiabetic effects with inhibition against PTPase enzyme activity [69]. Its flavonoids, tannins, glycosides, terpenoids, saponins, and polyphenols showed α-glucosidase inhibitory property. Like *Ceriops decandra*, this species also yielded 2-(2 methylphenyl)-1-phenyl-(z)-1-propene and 5S*,8S*,9S*,10R*,13S*)-18 hydroxy-16-nor-3-oxodolabr-4(18)-en-15-oic acid, which displayed both α-glucosidase and α-amylase inhibitory activities [68]. In the α-glucosidase enzyme inhibitory activity test, the extract of this species demonstrated a better effect with an IC50 value of 1.6 mg/mL compared to the extract of *Ceriops decandra*. On the other hand, in the α-amylase enzyme inhibitory activity test, this species showed a poor performance compared to *Ceriops decandra,* as this extract yielded an IC50 value of 4.0 mg/mL (Fig. **134**) [78].

Fig. (134). *Ceriops tagal.* Abu Hamas; Creative Commons Attribution-Share Alike 4.0 International license. https://commons.wikimedia.org/wiki/File:Ceriops_tagal_50589978.jpg

Excoecaria agallocha

Global distribution: Temperate and tropical Asia, Africa, and NW Australia.

Ecology: It is found in coastal regions (Fig. **135**).

Fig. (135). *Excoecaria agallocha.* Vengolis; Creative Commons Attribution-Share Alike 4.0 International license. https://commons.wikimedia.org/wiki/File:Excoecaria_agallocha_05938.JPG

Antidiabetics and their mechanisms of action: The flavonoids, tannins, polyphenols, and saponins present in the extracts of this species have been reported to help in the pancreatic secretion of insulin and uptake of glucose [69]. The oral use of the crude ethanolic extract of the leaves of this plant species at 500 mg/ kg body wt displayed potent hypoglycemic and anti-hyperglycemic activities in normal and alloxan-induced diabetic albino mice, respectively. Further, a single dose administration of 50% ethanolic extract (500mg/kg) of the leaves of this species in diabetic Wistar albino mice displayed a potent reduction in blood glucose level (FBGL) after 3 (39.1%) and 5h (41.4%) interval [79]. The effects of water and ethanolic extracts of the leaves of this species on blood glucose of diabetic Wistar albino mice are shown in Table **10**.

Table 10. Effects of the extracts of the leaves of *Excoecaria agallocha* on blood glucose level [79].

Exptl. Group	Blood Glucose Level at 0 hr (mg/ml)	Blood Glucose Level after 2 hr (mg/ml)
Control	107	98.4
Water extract	98.2	67.8
Ethanolic extract	89.6	51.2

Heritiera fomes

Global distribution: Indo-Pacific: India, Bangladesh, Malaysia, Myanmar, and Thailand.

Ecology: It grows in less saline habitats and on drier grounds that are inundated by the tides.

Antidiabetics and their mechanisms of action: The hot water extract of this species was found to help release insulin from BRIN BD11 cells. Further, the non-toxic concentrations of the hot water extract were found to stimulate insulin release from the isolated mouse islets and clonal pancreatic β-cells. Further, the oral administration of the extract at 250 mg/5ml/kg b.w to high-fat fed rats led to a significant improvement in plasma insulin responses and glucose tolerance besides inhibiting plasma DPP-IV activity. Furthermore, the extract reduced glucose absorption while increasing unabsorbed sucrose transit [80]. Its tannins, terpenoids, and flavonoids helped in the inhibition of glucose absorption in the gut and increased pancreatic secretion of insulin (Fig. **136**) [68].

Kandelia candel

Global distribution: It is found along the coasts of South Asia and SE Asia, from western India to Borneo.

Ecology: Its habitats include both saline and freshwater areas along the intertidal zonation.

Antidiabetics and their mechanisms of action: The flavonoids, triterpenoids, tannins, saponins, glycosides, and polyphenols present in the extracts of this species have shown antidiabetic activity [69]. The crude ethanolic extract of the leaves of this plant species showed significant antihyperglycemic activity. At a dose of 500 mg/kg, the SLM and STZ model experimental rats showed a lowering of blood sugar with percentage values of 35 and 20.5, respectively. Further, at 250 mg/kg dose, the corresponding values were 38 and 22%, respectively (Fig. **137**) [81].

Fig. (136). *Heritiera fomes.* Monster eagle ; Creative Commons Attribution-Share Alike 3.0 Unported license. https://commons.wikimedia.org/wiki/File:Sundarbans_02.jpg

Fig. (137). *Kandelia candel.* Shagil Kannur; Creative Commons Attribution-Share Alike 4.0 International license. https://commons.wikimedia.org/wiki/File:Mangroves_at_Muzhappilangad_021.jpg

Nypa fruticans

Global distribution: Its distribution is restricted to the tropical Indo-West Pacific region.

Ecology: It grows in coastlines and estuarine habitats. It is known to grow very well in the brackish water forest or along the river near its mouth.

Antidiabetics and their mechanisms of action: The alkaloids, sterols, glycosides, and tannins of this plant species exhibited antidiabetic by utilizing glucose [69]. Its bioactive compounds showed antidiabetic activity by utilizing glucose either by the mediation of enhanced insulin secretion or *via* the direct stimulation of glucose uptake [68]. The aqueous extract of this species showed antihyperglycaemic activity with a potent blood glucose lowering effect (56.6%) and enhanced serum insulin levels (79.8%) (Fig. **138**) [82].

Fig. (138). *Nypa fruticans*. GFDL; Creative Commons Attribution-Share Alike 3.0 Unported license. https://commons.wikimedia.org/wiki/File:Nypa_fruticans_Wurmb.jpg

Rhizophora annamalayana

Global distribution: Southeast Asia and Indonesia to western Pacific and northern Australia.

Ecology: It is commonly seen in mid-intertidal areas and intermediate and downstream estuarine areas.

Antidiabetics and their mechanisms of action: Its alkaloids, tannins, and steroids have shown antidiabetic effects by improving the level of insulin secretion and its action [68 - 69].

Rhizophora apiculata

Global distribution: Throughout SE Asia and the western Pacific islands.

Ecology: It is seen along the inter-tidal areas of creeks and channels in the sheltered parts of mangrove forests, nearer to estuarine areas (Fig. **139**).

Fig. (139). *Rhizophora apiculata.* Dinesh Valke from Thane, India; Creative Commons Attribution-Share Alike 2.0 Generic license. https://commons.m.wikimedia.org/wiki/File:Rhizophora_apiculata_Blume_%2853231938109%29.jpg

Antidiabetics and their mechanisms of action: Its tannin, phenols, steroids, and triterpenes showed antidiabetic effects by enhancing insulin secretion and its action, beta-cell protection, and insulin-mimetic activity [69]. The administration of the ethanol extract and dichloromethane and acid aqueous fractions of this species lowered glucose levels in the experimental rats of IDDM and NIDDM models, as shown in Table **11** [72].

Table 11. Effects of extracts of *Rhizophora apiculata* on blood glucose levels (mg/ml) in experimental rats [72].

Group	Extract & Conc	O day	7th day	14th day	21st day
NDDM	EE, 250mg.kg	745	555	523	193
NDDM	DCM, 250mg.kg	892	632	974	385
NDDM	AA, 250mg/kg	853	685	422	222
NDDM	Control	783	044	753	152
IDDM	EE, 250mg.kg	652	422	577	220

(Table 11) cont.....

Group	Extract & Conc	O day	7th day	14th day	21st day
IDDM	DCM, 250mg/kg	660	402	558	133
IDDM	AA, 250mg/kg	673	435	581	244
IDDM	Control	615	945	556	050

EE: Ethanol extract; DCM: Dichloromethane fraction: AA: Acid aqueous fraction.

Rhizohora mangle

Global distribution: Tropical and sub-tropical parts of America, Africa, and Fiji.

Ecology: It is found restricted to coastal, estuarine ecosystems (Fig. **140**).

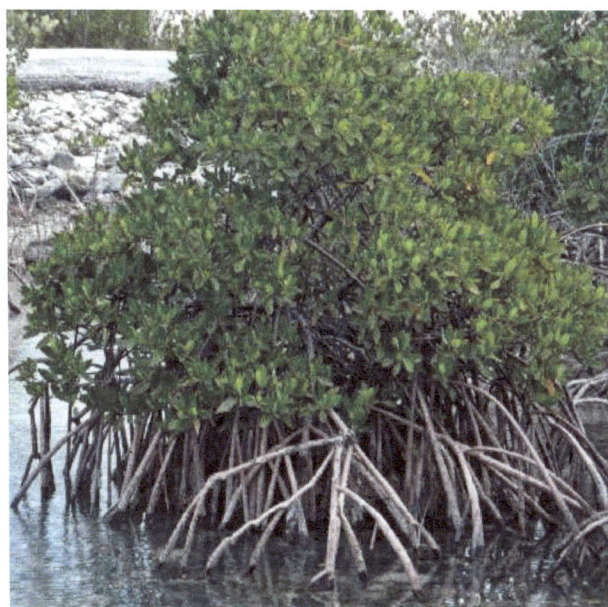

Fig. (140). *Rhizohora mangleames* St. John; Creative Commons Attribution 2.0 Generic license. https://commons.wikimedia.org/wiki/File:Rhizophora_mangle_%28red_mangrove%29_%28San_Salvador_Is land,_Bahamas%29_1_%2815784499775%29.jpg .

Antidiabetics and their mechanisms of action: Its tannins and triterpenes have been reported to yield antidiabetic properties [69]. Its extract showed the presence of compounds *viz.* cinchonains Ia and Ib, catechin-3-O-rhamnopyranoside, epicatechin, lyoniside, and nudiposide (Figs. **141-146**), the bioactivities of which are yet to be known. The oral administration of its extract twice a day over a period of 42 days was found to exert chronic hypoglycemic effects in streptozotocin–nicotinamide-induced hyperglycemic rats [83].

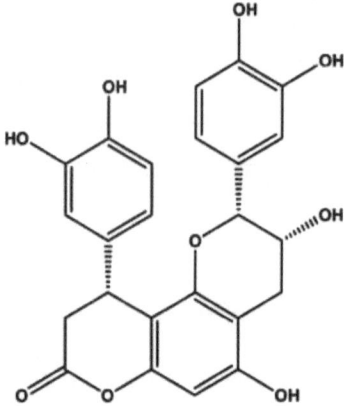

Fig. (141). Cinchonains Ia and Ib.

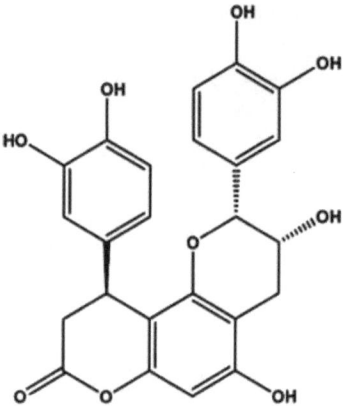

Fig. (142). Cinchonains Ia and Ib.

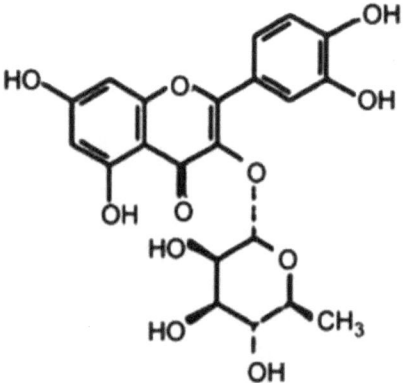

Fig. (143). Catechin-3-O-rhamnopyranoside.

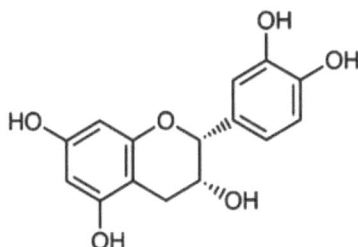

Fig. (144). Epicatechin.

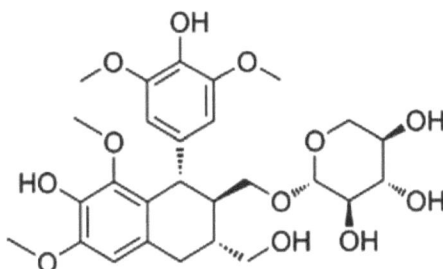

Fig. (145). Lyoniside.

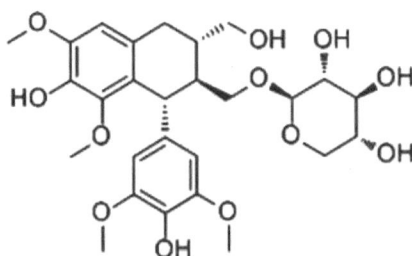

Fig. (146). Nudiposide.

Rhizophora mucronata

Global distribution: Indo-Pacific.

Ecology: It is seen on the banks of rivers and the edge of the sea.

Antidiabetics and their mechanisms of action: The ethyl acetate extracts of this species exhibited antidiabetic effects by significantly reducing blood glucose levels in mice by 57.64% after 24 hours of oral administration [84]. Its tannin, triterpenes, steroids, and phenolic compounds improved the level of insulin secretion and its action, α-glucosidase inhibitory activity, and insulin-mimetic activity [69]. The oral use of the extract mixture of this plant species and *Avicennia marina* to diabetic groups yielded a significant increase in the serum insulin levels compared to the untreated diabetic rats [73]. Further, it was also reported that the administration of the extracts of this species potently reduced diabetes complications with alpha-glucosidase and alpha-amylase inhibitory activities with IC50 values of 24.4 and 41.1ugML (Fig. **147**) [85].

Fig. (147). *Rhizophora mucronata.* Shagil Kannur; Creative Commons Attribution-Share Alike 3.0 Unported license.   https://commons.wikimedia.org/wiki/File:Rhizophora_mucronata_with_propagule_at_ Muzhappilangad.jpg

Rhizophora racemosa

Global distribution: Pacific coast of Central and South America.

Ecology: It is mostly seen in highly brackish habitats with rich freshwater inputs.

Antidiabetics and their mechanisms of action: The extracts of this species containing tannins and triterpenes have shown antidiabetic properties [69]. The administration of its extracts at 400 mg/kg reduced the average blood glucose level from 110.6 to 93.2 mg/dL in experimental mice (Fig. **148**) [86].

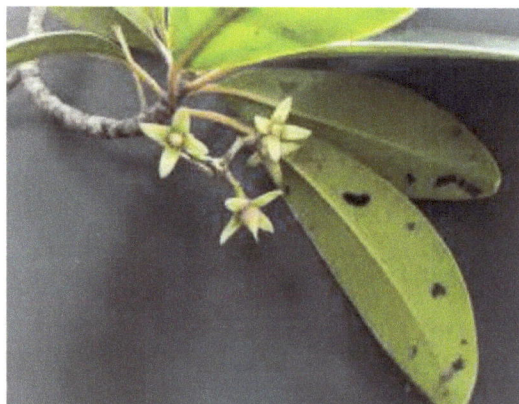

Fig. (148). Rhizophora racemosa. Sommerluk; Creative Commons CC0 1.0 Universal Public Domain Dedication. Creative Commons CC0 1.0 Universal Public Domain Dedication.

Rhizophora stylosa

Global distribution: Pacific: Japan, China, Taiwan, Cambodia, Vietnam, and Australia.

Ecology: It is found in the intertidal wetland zone of 0–6 m elevation with variable rainfall.

Antidiabetics and their mechanisms of action: Its inositols and steroids have been reported to possess antidiabetic properties (Fig. **149**) [69].

Fig. (149). *Rhizophora stylosa.* Kevin Thiele from Perth, Australia; Creative Commons Attribution 2.0 Generic license. https://en.m.wikipedia.org/wiki/File:Rhizophora_stylosa_-_Flickr_-_Kevin_Thiele.jpg

Rhizophora sp.

Antidiabetics and their mechanisms of action: The ethanol extract of the leaves of this plant species displayed excellent antidiabetic activity and lowered blood glucose levels [84].

Sonneratia alba

Global distribution: SE Asia.

Ecology: This tree sporadically occurs on tidal channels.

Antidiabetics and their mechanisms of action: The tannins, polysaccharides, and phenolic compounds available in the extracts of this species yielded antidiabetic properties by modifying the glucose pathway (Fig. **150**) [69].

Fig. (150). *Sonneratia alba.* Ton Rulkens from Mozambique, Creative Commons Attribution-Share Alike 2.0 Generic license. https://commons.wikimedia.org/wiki/File:Sonneratia_alba_-_fruit_(8349980264).jpg

Sonneratia apetala

Global distribution: It is restricted to southern India, Sri Lanka, South Andaman Isl., Myanmar, and Vietnam.

Ecology: It is seen along the intertidal estuarine regions of mangrove forests.

Antidiabetics and their mechanisms of action: Its triterpenes steroids and flavonoids alkaloids have shown antidiabetic effects by enhancing the insulin-releasing activity and transport of blood glucose to the peripheral tissue (Fig. **151**) [69].

Fig. (151). *Sonneratia apetala.* Dinesh Valke from Thane, India; Creative Commons Attribution-Share Alike 2.0 Generic license. https://commons.wikimedia.org/wiki/File:Sonneratia_apetala_%285355156316%29.jpg

Sonneratia caseolaris

Global distribution: Bangladesh, Malaysia, Myanmar, Philippines, Thailand, Northeast Australia, Papua New Guinea, Brunei Darussalam, Cambodia, China, Indonesia, and India.

Ecology: It lives in the mangrove forests in tidal areas with mud banks. Occasionally, it is also found growing in freshwater.

Antidiabetics and their mechanisms of action: The steroid and glycosides present in the extracts of this plant species showed intestinal α-glucosidase inhibitory effects and improved the pancreatic secretion of insulin [69]. The methanol extract (85%) of the mature fruits of this plant species decreased blood glucose levels in experimental mice (Fig. **152**) [87].

Fig. (152). *Sonneratia caseolaris.* Renjusplace ; Creative Commons Attribution-Share Alike 4.0 International license. https://commons.wikimedia.org/wiki/File:Sonneratia_caseolaris,_Apple_Mangrove

Sonneratia ovata

Global distribution: China, Palau, New Guinea, and Australia.

Ecology: It is seen on tidal river banks and muddy soils influenced by spring tides.

Antidiabetics and their mechanisms of action: The steroid content of the extract of this plant species displayed antidiabetic properties (Fig. **153**) [69].

Fig. (153). *Sonneratia ovate.* Abu Hamas; Creative Commons Attribution-Share Alike 4.0 International license. https://commons.wikimedia.org/wiki/File:Sonneratia_ovata.jpg

Xylocarpus granatum

Global distribution: Africa, Asia, Pacific Islands, and Australasia.

Ecology: It is found in the upper intertidal areas and estuaries.

Antidiabetics and their mechanisms of action: Its alkaloids, steroids, triterpenes, tannins, flavonoids, and alkaloids have shown antidiabetic effects by stimulating the β- cells and thereby improving insulin sensitivity to glucose [69]. Further, a limonoid derivative Xyloccensin-I derived from the bark and stem of this species yielded potent α-glucosidase and α-amylase inhibitory activities with IC50s of 0.16 and 0.25 mg/ml, respectively (Fig. **154**) [68].

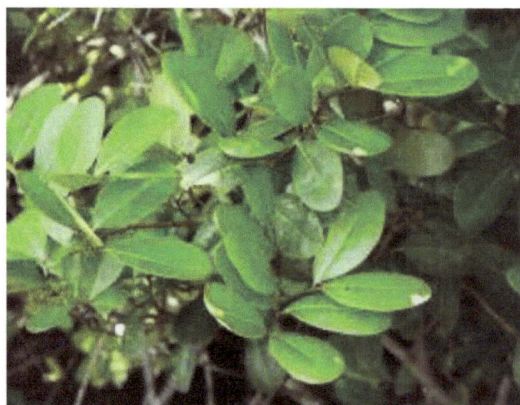

Fig. (154). *Xylocarpus granatum.* Dinesh Valke from Thane, India; Creative Commons Attribution-Share Alike 2.0 Generic license. https://commons.wikimedia.org/wiki/File:Xylocarpus_granatum_%285929385514%29.jpg

Xylocarpus moluccensis

Global distribution: W. Pacific; Somalia to N. Mozambique.

Ecology: It is seen sporadically in the interior elevated areas of mangrove forests (Fig. **155**).

Antidiabetics and their mechanisms of action: The alkaloids, steroids, tannins, proanthocyanidins, and triterpenes present in the extracts of this species displayed antidiabetic effects with insulin secretagogue and resistance reversal activity [69]. Among its two tetrahydroxyterpenoids compounds, namely xyloccensin E (Fig. **156**) and ruageanin D, and one flavonoid compound, namely 3,5,7,3',4'-pentahidroxyflavan (catechin) (Fig. **157**), xyloccensin E and catechin exhibited α-glucosidase inhibitory activity with IC50 values of 118.6 and 55.2 µg/mL, respectively [88].

Fig. (155). *Xylocarpus.moluccensis.* Wibowo Djatmiko (Wie146); Creative Commons Attribution-Share Alike Attribution-Share Alike 4.0 International, 3.0 Unported, 2.5 Generic, 2.0 Generic and 1.0 Generic license. https://commons.wikimedia.org/wiki/File:Xyloc_moluc_191103-2923_skd.jpg

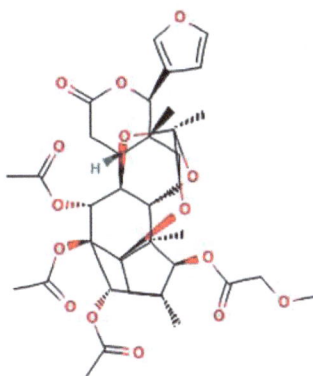

Fig. (156). Xyloccensin E.

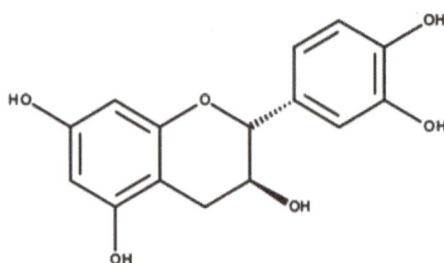

Fig. (157). Catechin.

Coastal Plants

Cerbera manghas

Global distribution: Tanzania; W. Indian Ocean to the Pacific Ocean.

Ecology: It is normally associated with water, and it is found along rivers or streams (Fig. **158**).

Fig. (158). *Cerbera manghas,*WingkLEE; Creative Commons Attribution-Share Alike 3.0 Unported license. https://commons.wikimedia.org/wiki/File:Cerbera_manghas_flower.jpg

Antidiabetics and their mechanisms of action: The macrolide des-O methyllasiodiplodin (Fig. **159**) derived from this species has shown antidiabetic effects by potently reducing the blood glucose levels and HbA1c in mice (Chellappan *et al.*, 2023).

Clerodendrum inerme-derived endophytic fungus *Trichoderma* sp. strain 307

Global distribution: Tropical Asia and the Pacific Ocean.

Ecology: It is mainly seen in open coastal areas, *i.e.* along the beach, in mangroves, along tidal rivers along rivers, and in swamps.

Antidiabetics and their mechanisms of action: The presence of alkaloids, glycosides, saponins, phenolic compounds, proteins, phytosterols, and flavonoids in the ethanol extract of leaves of this plant species showed significant hypoglycemic activity in alloxan-induced diabetic rats when they were given the extract at 200 mg/kg (Dhanabal *et al.*, 2008) (Fig. **160**).

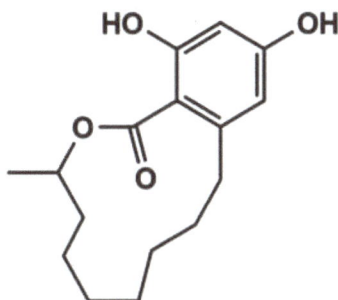

Fig. (159). des-O-methyllasiodiplodin.

Fig. (160). *Clerodendruminerme*. Vengolis; Creative Commons Attribution-Share Alike 3.0 Unported license. https://commons.wikimedia.org/wiki/File:Volkameria_inermis_08611.jpg

CONCLUSION

It is worth mentioning here that marine macroalgae have been identified as a rich and potential source of promising bioactive compounds of therapeutic value. Although these algae are marketed as functional foods of nutraceutical importance, no antidiabetic product lines have yet demonstrated from these algae that they are useful and pharmaceutically feasible. Consumer interest in employing natural bioactive compounds as diabetic medications has recently increased, and algae's numerous biological processes have the potential to boost their value as health-beneficial ingredients in pharmaceutical and functional food industries. Recent research investigations have proved that the active components derived from the different species of marine algae are of great use for the treatment of diabetes. All such components offer vast scope for the development of new antidiabetic drugs with nil or fewer side effects. Therefore, intensive and collaborative research between marine and pharmaceutical scientists is the need of the hour.

CHAPTER 5

Antidiabetic Properties of Marine Invertebrates

Abstract: The chemical diversity (terpenes, alkaloids, steroids, peptides, *etc.*) of marine invertebrates such as sponges, cnidarians, annelid worms, crustaceans, mollusks, and echinoderms and their mechanisms of action has been dealt with in this chapter. The mechanisms of action of these natural products, along with their potential in the discovery of novel drugs, are also given in detail.

Keywords: Chemical diversity, Marine invertebrates, Mechanisms of action, Therapeutic potential.

INTRODUCTION

The marine invertebrates such as sponges, cnidarians, annelid worms, crustaceans, mollusks, and echinoderms have yielded a substantial diversity of natural products, including terpenes, alkaloids, steroids, aliphatic hydrocarbons, carbohydrates, amino acids, peptides, *etc.*, which offer immense scope in the industrial applications as agricultural products, pharmaceutics, antibiotics, nutraceutics, cosmetics, biomaterials, *etc.*

Sponges

The sponges with about 8000 described species are found to be widely distributed in marine and freshwater environments. These sessile organisms are considered to be a potential source of an enormous diversity of therapeutic compounds. The pharmaceutical interest among sponges started in the 1950s with the investigation of the therapeutically important nucleosides *viz.* spongothymidine and spongouridine from the species *Cryptotethya crypta*. These nucleosides were largely responsible for the synthesis of ara-A, an antiviral drug, and ara-C, the first marine-derived anticancer agent. Further, marine sponges have been reported to be the most important producer of novel bioactive compounds, and about 5000 compounds are believed to be extracted every year from these organisms. It is worth mentioning here that some of these secondary metabolites are in the process of a clinical and pre-clinical trial (*e.g.*, as anticancer or anti-inflammatory agents) [89]. However intensive research on the antidiabetics of these organisms is

Santhanam Ramesh, Ramasamy Santhanam & Arumugam Uma

lacking, although a wide chemical diversity of antidiabetic compounds, such as sesquiterpenes, quinones, and hydroquinones, have been derived from the marine sponge species, such as *Dysidea villosa*, *Callyspongia truncata,* and *Lamellodysidea herbacea* [90].

Agelas mauritiana

Global distribution: Tropical seas of Indo-West Pacific (Fig. **1**). Ecology: This sessile species lives at a depth range of 0 - 100 m.

Fig. (1). *Agelas* sp. Image credit: Esculapio ; Creative Commons Attribution 3.0 Unported license. https://commons.wikimedia.org/wiki/File:Agelas_oroides_Capo_Gallo_025.JPG

Antidiabetics and their mechanisms of action: α-galactosylceramide (αGalCer, KRN7000) (Fig. **2**) derived from this species is believed to protect beta-pancreatic cells and activate natural killer cells [25]. Natural Killer (NK) cells are cytotoxic lymphocytes that engage in innate immunity to remove pathogens and cancer cells in patients with type 2 diabetes mellitus.

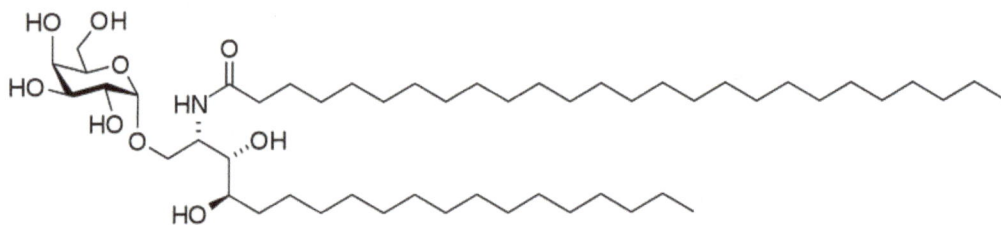

Fig. (2). α-galactosylceramide.

Agelas nakamurai

Global distribution: Tropical and subtropical West Indies, Mediterranean Sea, Red

Sea, and Indian Ocean.

Ecology: It normally resides in the shallow seas up to a depth of about 30 m.

Antidiabetics and their mechanisms of action: An N-methyladenine-derived sesquiterpene *viz.* agelasine G (Fig. **3**) isolated from this species has shown Protein Tyrosine Phosphatase 1B (PTP1B) inhibiting activity with an IC_{50} value of 15 µM [2]. PTP1B is mostly involved in the negative regulation of signaling mediated by the insulin and leptin receptors. This enzyme is, therefore, known to play an important role in the development of diseases associated with insulin resistance, such as obesity and diabetes.

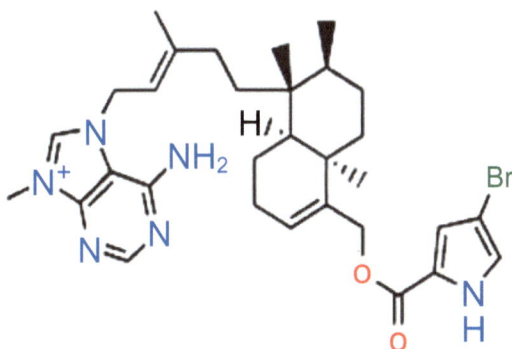

Fig. (3). Agelasine G.

Amphimedon viridis (= Haliclona viridis)

Global distribution: Caribbean, Bermuda, North Carolina, West central Pacific, and Indian Ocean.

Ecology: It commonly inhabits shallow reefs and seagrass beds. Antidiabetics and their mechanisms of action:

An ethanol extract of this species has shown a significant hypoglycemic effect, lasting for more than 8 h after the administration of single oral doses of 200 - 500 mg/kg to experimental mice [91] (Fig. **4**).

Axinyssa **sp.**

Global distribution: It is a common species of French Polynesia: Tuamotu Islands. Ecology: It is often collected from the lagoon on pinnacles.

Antidiabetics and their mechanisms of action: Two unidentified organic compounds (Fig. **5**) derived from this sponge species have shown antidiabetic activity by inhibiting the PTP1B enzyme with IC_{50} values of 1.9 μM and 17 μM, respectively [2].

Fig. (4). *Amphimedon viridis.* Image credit: Soledade, G.O. *et al.* ; Creative Commons Attribution 4.0 International license. https://commons.wikimedia.org/wiki/File:Amphimedon_viridis_%2810.1590-23--2936e2017027%29_Figure_1_%28cropped%29.jpg

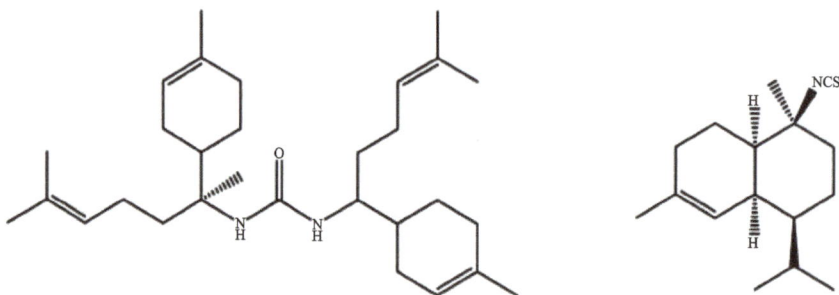

Fig. (5). Unidentified organic compounds.

Callyspongia lindgreni (= Callyspongia truncata)

Global distribution: It is found distributed in the tropical Central and Western Pacific, as well as Indian, West Atlantic, and East Pacific oceans.

Ecology: This benthic species is commonly found living attached to rocks or other substrates (Fig. **6**).

Antidiabetics and their mechanisms of action: A C_{32} polyacetylenic acid *viz.* callyspongynic acid (Fig. **7**) and an acetylenic acid, corticatic acid (Fig. **8**) isolated from this species, inhibited α-glucosidase with IC_{50} values of 0.25 μg/mL and 0.16 μg/mL, respectively [2].

Fig. (6). *Callyspongia lindgreni* . Image credit: Dept. of Natural Product Chemistry, Graduate School of Pharmaceutical Sciences, Osaka University; Creative Commons CC0 License ; https://commons.wikimedia.org/wiki/File:Callyspongia.png.

Fig. (7). Callyspongynic acid.

Fig. (8). Corticatic acid.

Clathria (Clathria) prolifera

Global distribution: It is native to the western Atlantic Ocean, from Prince Edward Island to Florida and Mexico.

Ecology: It commonly occurs in shallow bays and harbors in dock fouling and on the undersides of subtidal and low intertidal rocks (Fig. **9**).

Fig. (9). *Clathria (Clathria) prolifera.* Image credit: Andrew C ; Creative Commons Attribution 2.0 Generic license.
https://commons.wikimedia.org/wiki/File:Red_Beard_Sponge_(Clathria_prolifera)_(16133310360).jpg

Antidiabetics and their mechanisms of action: A tricyclic spiroketal compound clathriketal (7-(hydroxymethyl)-13-methoxy-3,11-dimethy-4-oxo-octahydrospiro [chromene-9,13-pyran]-11-yl propionate) (Fig. **10**) derived from this species has shown significant anti-hyperglycemic property by inhibiting serine protease dipeptidyl peptidase-IV with an IC_{50} value of 0.37 mM. Further, this spiroketal compound was found to display significant inhibitory activities against carbolytic enzymes α-glucosidase and α-amylase with IC_{50} values of 0.43 mM and 0.41 mM, respectively. Furthermore, clathriketal showed promising anti- hyperglycemic activity by significantly inhibiting oxidants 2, 2'-azino-bis- 3-ethylbenzthiazolin- -6-sulfonic acid and 2, 2-diphenyl-1-picrylhydrazyl ($IC_{50} \sim 1.2$ mM) [92].

Dysidea avara

Global distribution: Tropical Atlantic and Eastern Central Pacific Oceans (Fig. **11**). Ecology: This sessile species often dwells in the rocky sublittoral areas.

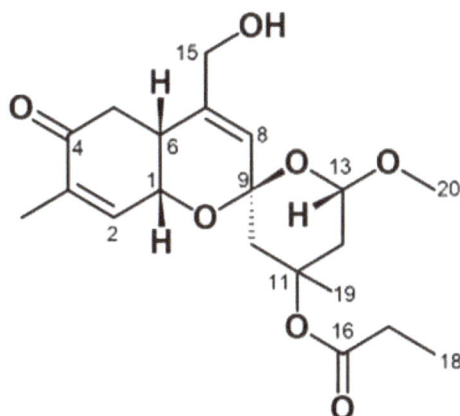

Fig. (10). Clathriketal.

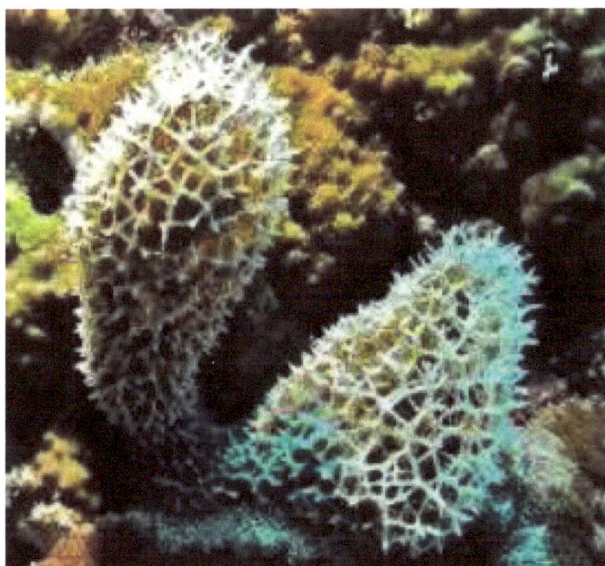

Fig. (11). *Dysidea* sp. Image credit: Bernard DUPONT from FRANCE; Creative Commons Attribution-Share Alike 2.0 Generic license. https://commons.wikimedia.org/wiki/File:Dysideid_Sponge_%28Dysidea_sp .%29_showing_%22skeleton%2 2_%288481640074%29.jpg.

Antidiabetics and their mechanisms of action: The sesquiterpene quinones *viz*. dysidavarones A and D (Fig. **12**) derived from this species have shown PTP1B inhibitory activity with IC_{50} values of 9.98 μM and 21.6 μM, respectively. Further, its avarone, 3-(methylamino) avarone, 4-(methylamino)avarone, and avarol (Fig. **13**) have also been reported to show similar activity with IC50 values of 6.7, 15.2, 21.6, and 42.2 μM, respectively. Furthermore, its avarone and avarol have also shown potent inhibitory activities against α-glucosidase with percentage values of 86.2 and 78.9, respectively, at 10 μM [2-24].

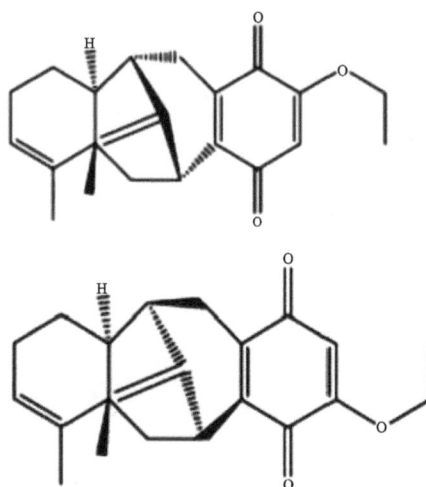

Fig. (12). Dysidavarone A and D.

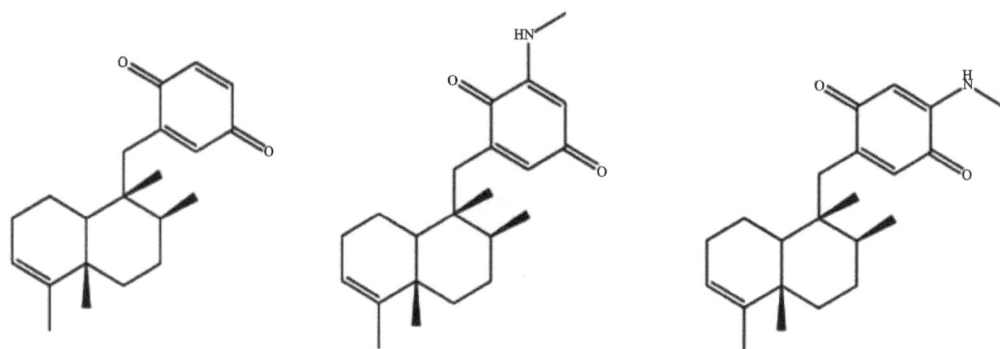

Avarone.   3-(methylamino)avarone.   4-(methylamino)avarone

Fig. (13). Avarone, 3-(methylamino)avarone, 4-(methylamino)avarone.

Dysidea cinerea

Global distribution: It is distributed in the tropical Western Pacific, including Indonesia and Hong Kong.

Ecology: This sessile species is commonly seen in the shallow seashore areas.

Antidiabetics and their mechanisms of action: Its sesquiterpenes *viz.* cinerol A-C and F (Figs. **14, 15**) have shown PTP1B inhibitory activity with IC50 values of 3.9, 0.9, 8.8, and 0.7 µM, respectively [2].

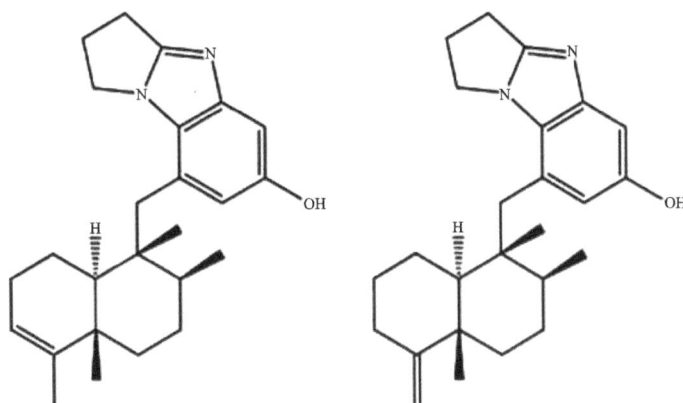

Fig. (14). Cinerol A,B.

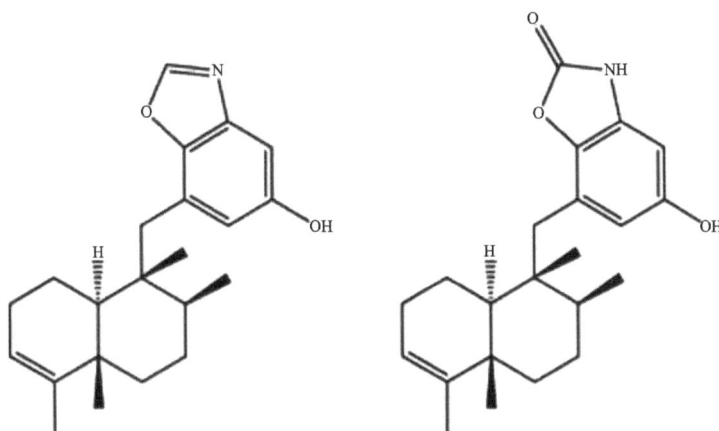

Fig. (15). Cinerol C,F.

Dysidea frondosa

Global distribution: It is distributed in the tropical Central Pacific, including French Polynesia and New Caledonia.

Ecology: This sessile species is commonly seen on dead corals and at a depth range of 0 - 100 m.

Antidiabetics and their mechanisms of action: Two sesquiterpenes *viz.* frondoplysin A,B (Fig. **16**) derived from this sponge have shown PTP1B inhibitory activities with IC_{50} values of 0.39 and 0.65 μM, respectively [2].

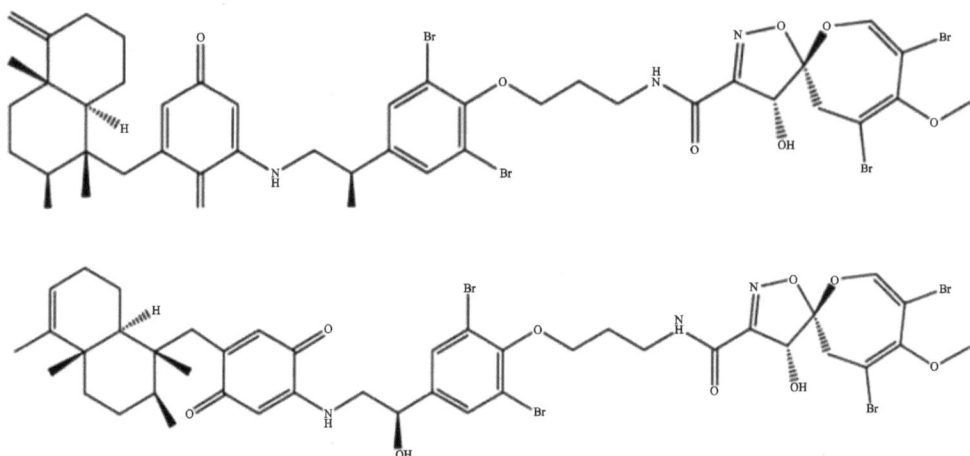

Fig. (16). Frondoplysin A, B.

Dysidea septosa

Global distribution: It is found in the Red Sea, Australia, Yap State, and the Philippines

Ecology: It is a nearshore species

Antidiabetics and their mechanisms of action: Three sesquiterpenes *viz.* hydroxybutenolide, microcionin, and dihydropallescensin-2 (Fig. **17**) derived from this sponge species have shown PTP1B inhibitory activities with IC_{50} values of 8.8, 11.6, and 6.8 µg/mL, respectively. Further, a 6/8-bicyclic furanosesquiterpene Nakafuran-8 (Fig. **18**) of this species has demonstrated similar activity with an IC_{50} value of 1.9 µg/ml [2].

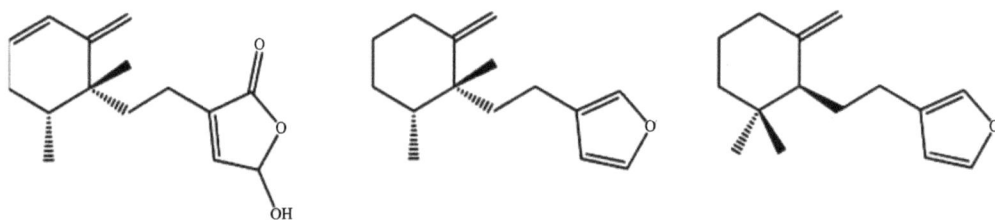

Hydroxybutenolide   Microcionin   Dihydropallescensin-2

Fig. (17). Hydroxybutenolide, Microcionin, Dihydropallescensin-2.

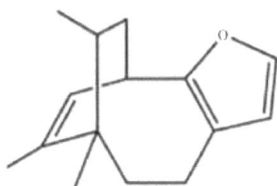

Fig. (18). Nakafuran-8.

Dysidea villosa

Global distribution: It has been reported from the South China Sea, the Red Sea, Australia, Yap State, and the Philippines.

Ecology: This benthic and sessile species is found mainly in the nearshore areas.

Antidiabetics and their mechanisms of action: The sesquiterpene quinone, dysidine (Fig. **19**), derived from this species, demonstrated inhibition activity against PTP1B with an IC_{50} value of 1.5 μM [2]. The dysidine of this species showed inhibitory activity of PTP1B with an IC50 value of 6.7 μM. [24]. This species also yielded a sequiterpene quinone *viz*. 21-dehydroxybolinaquinone and its two analogs, bolinaquinone (Fig. **20**) and dysidine. Among these compounds, dysidine and bolinaquinone exhibited potent PTP1B inhibitory effects with IC50s of 6.70 and 5.45 μM, respectively. However, its 21-dehydroxybolinaquinone showed only moderate activity (IC $_{50}$ 39.50 μM) [6].

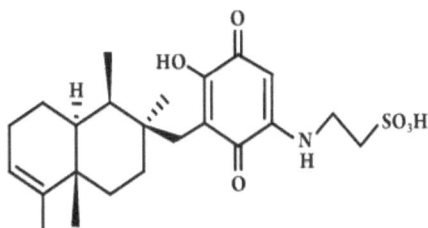

Fig. (19). Dysidine.

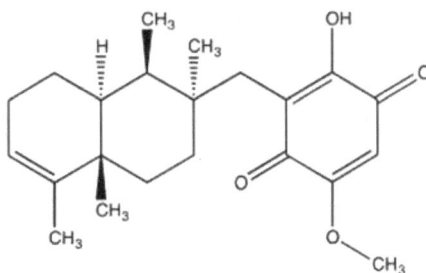

Fig. (20). Bolinaquinone.

Dysidea sp.

Antidiabetics and their mechanisms of action: A total of 9 sesquiterpene hydroquinones derived from this unidentified *Dysidea* sp have shown PTP1B inhibitory activity, as given in Table **1**.

Table 1. PTP1B inhibitory activity of sesquiterpene hydroquinones and lactone from *Dysidea* sp [2].

Compound	IC$_{50}$ (µM)
Avapyran (Fig. **21**)	11
Avarol (Fig. **23**)	12
Neoavarol (Fig. **24**)	35% inhibition at 32 µM
3′-aminoavarone (Fig. **22**)	18
17-*O*-acetylavarol (Fig. **23**)	9.5
Acetylneoavarol (Fig. **24**)	6.5
20-*O*-acetylavarol (Fig. **23**)	10
20-*O*-acetylneoavarol (Fig. **24**)	8.6
O-methylnakafuran-8 lactone*(Fig. **25**)	1.6

* Lactone (cyclic organic ester)

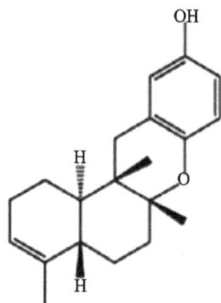

Fig. (21). Avapyran.

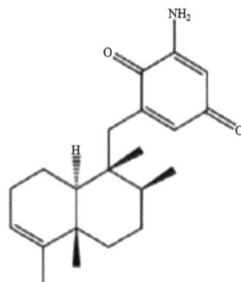

Fig. (22). 3′-aminoavarone.

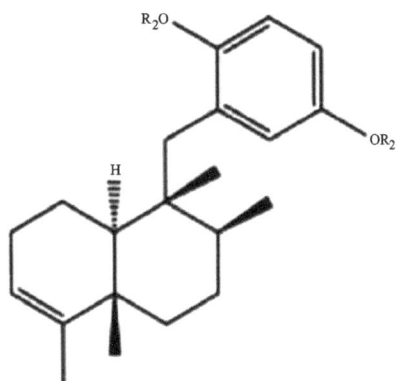

Fig. (23). Avarol (R1=H; R2=H) 17-O-acetylavarol (R1=Ac; R2=H). 20-O-acetylavarol (R1=H; R2=Ac).

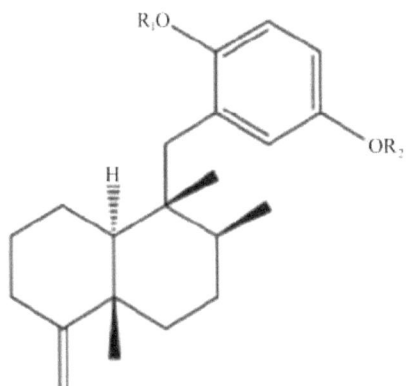

Fig. (24). Neoavarol (R1=H; R2=H) 17-O-acetylneoavarol (R1=Ac; R2=H) 20-O-acetylneoavarol (R1=H; R2=Ac).

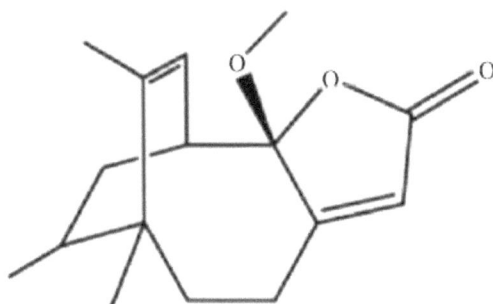

Fig. (25). O-methylnakafuran-8 lactone.

Euryspongia sp.

Global distribution: It has been found distributed in French Polynesia: Society islands and Windward islands.

Ecology: It dwells in the outer reef slope areas.

Antidiabetics and their mechanisms of action: The sesquiterpenes, euryspongin A-C (Figs. **26, 27**) and dehydroeuryspongin A (Fig. **28**), have been isolated from this species. Of these compounds, euryspongin A-C did not show any inhibition effect against PTP1B. On the other hand, the dehydro derivative of euryspongin A *viz*. dehydroeuryspongin A was found to exhibit inhibitory activity with an IC_{50} value of 3.6 μM [24, 25, 93].

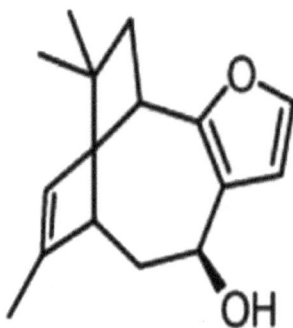

Fig. (26). Euryspongin A. Euryspongin B (R1=H).

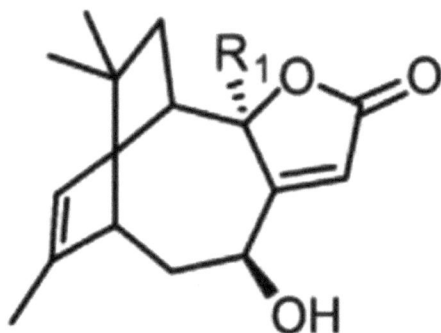

Fig. (27). Euryspongin C (R1=OCH3).

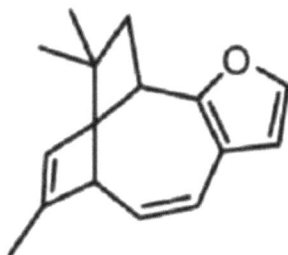

Fig. (28). Dehydroeuryspongin A.

Fascaplysinopsis reticulata

Global distribution: Tropical Western Central Pacific: Indonesia and New Caledonia.

Ecology: This sessile species has a depth range of 0 - 100 m.

Antidiabetics and their mechanisms of action: Two bromo-spiroalkaloids (Fig. **29**) derived from this species have been reported to display PTP1B inhibition activity with IC_{50} values of 7.67 and 11.25 μM, respectively [2].

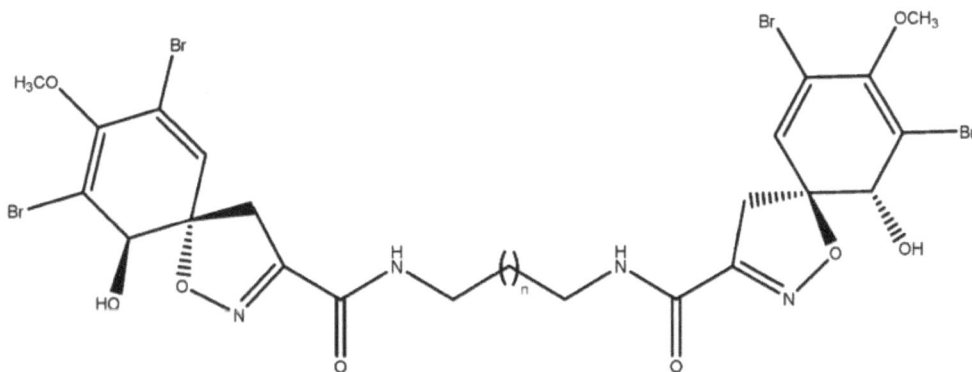

Fig. (29). Bromo-spiroalkaloids (1&2) (n=2; n=3).

Gelliodes pumila (= Sigmadocia pumila)

Global distribution: It is distributed in the Caribbean and eastern Pacific.

Ecology: It is mainly restricted to shallow-water fouling communities (*i.e.* floating docks or pier pilings) of harbors. It is also seen on the roots of the red mangrove *viz.* Rhizophora mangle.

Antidiabetics and their mechanisms of action: At a dosage of 250 mg/kg body, the extracts of this sponge species have been reported to show a little but insignificant lowering in blood glucose post sucrose load in experimental rats [94].

Halichondria (Halichondria) panicea

Global distribution: North Atlantic and Mediterranean Sea.

Ecology: It is found distributed from the intertidal zone to depths of more than 500 m (Fig. **30**).

Fig. (30). *Halichondria (Halichondria) panacea.* Image credit: Minette Layne ; Creative Commons Attribution-Share Alike 2.0 Generic license.; https://commons.wikimedia.org/wiki/File:Halichondria_panicea.jpg.

Antidiabetics and their mechanisms of action: The polyacetylene compounds isopetrosynol, petrosynol, adociacetylene D,and (5*R*)-3,15,27 triacontatrie- e-1,29-diyn-5-ol (Figs. **31-34**) isolated from this sponge species have shown inhibitory effects on PTP1B with IC50 values of 8.2, 28.9, 7.8, and 12.2 μM, respectively. Further, its petrosynol was also found to possess an inhibitory effect against α-glucosidase with an IC_{50} value of 4.08 μg/ml [2, 95].

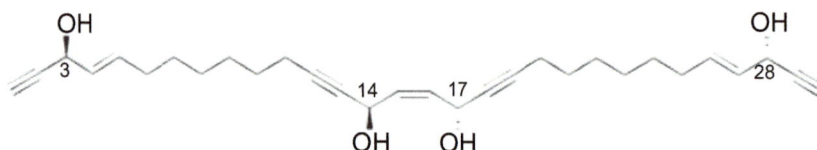

Fig. (31). Isopetrosynol.

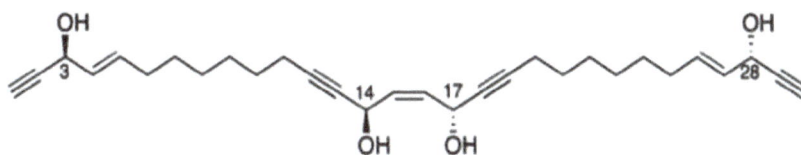

Fig. (32). Petrosynol.

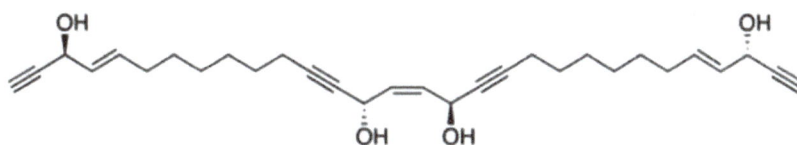

Fig. (33). Adociacetylene D.

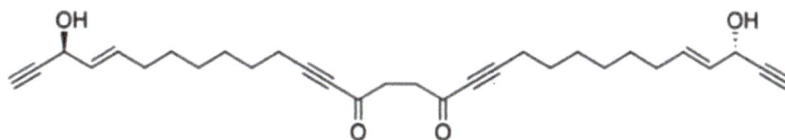

Fig. (34). 5*R*)-3,15,27-triacontatriene-1,29-diyn-5-ol.

Haliclona sp.

Global distribution: It is a common species of the French Polynesia: Society and Tuamotu islands (Fig.**35**).

Ecology: It is often seen in the outer reef slope areas.

Fig. (35). *Haliclona* sp. Image credit: Chaloklum Diving ; Creative Commons Attribution 3.0 Unported license. https://commons.wikimedia.org/wiki/File:Haliclona_sp,_Tailandia_2.jpg.

Antidiabetics and their mechanisms of action: The extracted metabolites of this sponge species have been reported to possess alpha-glucosidase and amylase, as shown in Table **2**.

Table 2. Alpha-glucosidase and amylase inhibitory activities (IC50, ug/ml) of *Haliclona* sp. [96].

Isolate	Alpha-glucosidase Inhibitory Activity (IC50, ug/ml)	Amylase Inhibitory Activity (IC50, ug/ml)
QH 26	177.6	9.0
QH 30	453.8	25.0
QH 36	153.5	15.6
QH 37	441.0	19.3
QN 45	341.8	21.5
QN 63	495.4	201.9

Hemimycale arabica

Global distribution: It is commonly found in Ras Um Sid, Egypt.

Ecology: This shallow-water species is often found growing on the tips of live or dead corals of the species *Millepora dichotoma*. It is known to harbor sabellid tube worms.

Antidiabetics and their mechanisms of action: Phenylmethylene hydantoin (Fig. **36**) derived from this species has shown GSK-3β inhibition and liver glycogen enhancement in experimental rats [8].

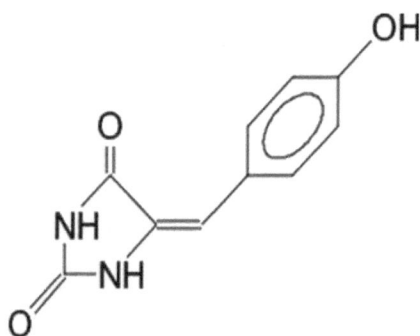

Fig. (36). 5-(4-Hydroxy-benzylidene)-hydantoin.

Hemimycale columella

Global distribution: It is found widely distributed across the Atlanto-Mediterranean region.

Ecology: The species is often found attached to boulders and bedrock in silt-free littoral areas (Fig. **37**).

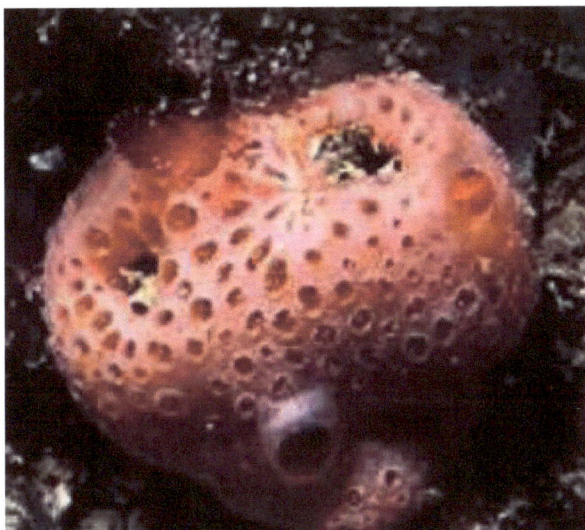

Fig. (37). *Hemimycale columella.* Image credit: Parent Géry ; e public domain .; https://commons.wikimedia.org/wiki/File:Hemimycale_columella_1_(Bowerbank,_1874).jpg.

Antidiabetics and their mechanisms of action: The water, butanol, and dichloromethane extracts of this species have yielded several glycosides with PTP1B inhibition activity. The different compounds isolated and their activities are given in Table **3**.

Table 3. Glycosides of *Hemimycale columella* and their PTP1B inhibition activity [2].

Compound	Extract	PTP1B inhibition (%)
2,3-O-hexahydroxydiphenoyl-(α/β)-glucose (Fig. **38**)	Water	80
Gentisic acid 2-O-β-glucoside(Fig. **39**)	Water	72.4
Quercetin-3-O-β-glucopyranoside (fig. **40**)	Butanol	32.4
Kaempferol 3-O-β-glucopyranoside (Fig. **40**)	Butanol	59.2
Isorhamnetin 3-O-β-glucopyranoside (Fig. **40**)	Butanol	25.3
Gallic acid (Fig. **41**)	Dichloromethan	76.2
Gallic acid-3-methyl ether (Fig. **41**)	Dichloromethan	52

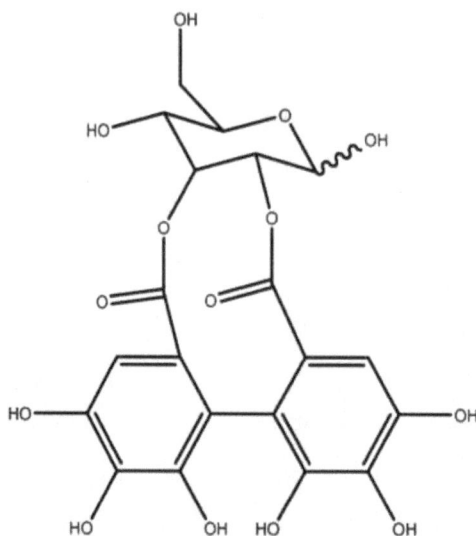

Fig. (38). 2,3-O-hexahydroxydiphenoyl-(α/β)-glucose.

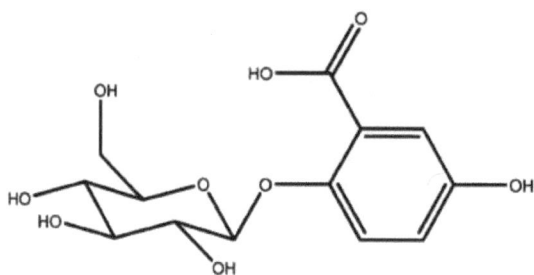

Fig. (39). Gentisic acid 2-O-β-glucoside.

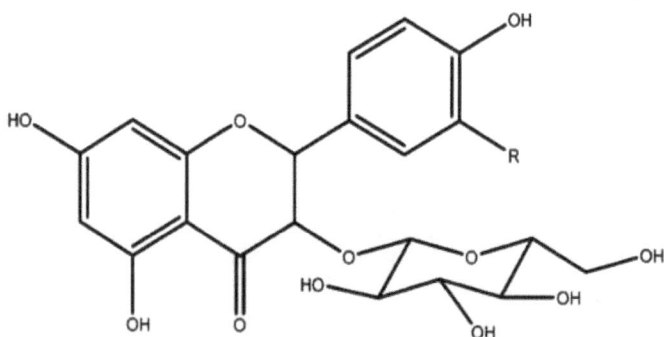

Fig. (40). Quercetin-3-O-β-glucopyranoside (R=OH), Kaempferol 3-O-β-glucopyranoside (R=H), Isorhamnetin 3-O-β-glucopyranoside (R= OCH3).

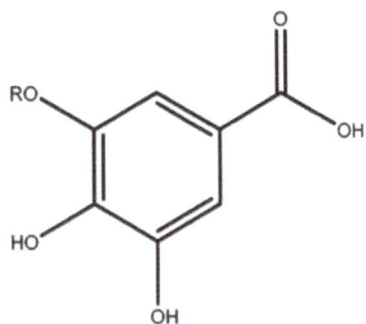

Fig. (41). Gallic acid (R=H) Gallic acid-3-methyl ether (R= CH3).

Hippospongia lachne

Global distribution: It has been reported from the tropical Indo-West Pacific and Western Central Atlantic (Fig. **42**).

Ecology: This sessile and non-migratory species has a depth range of 5 - 15 m.

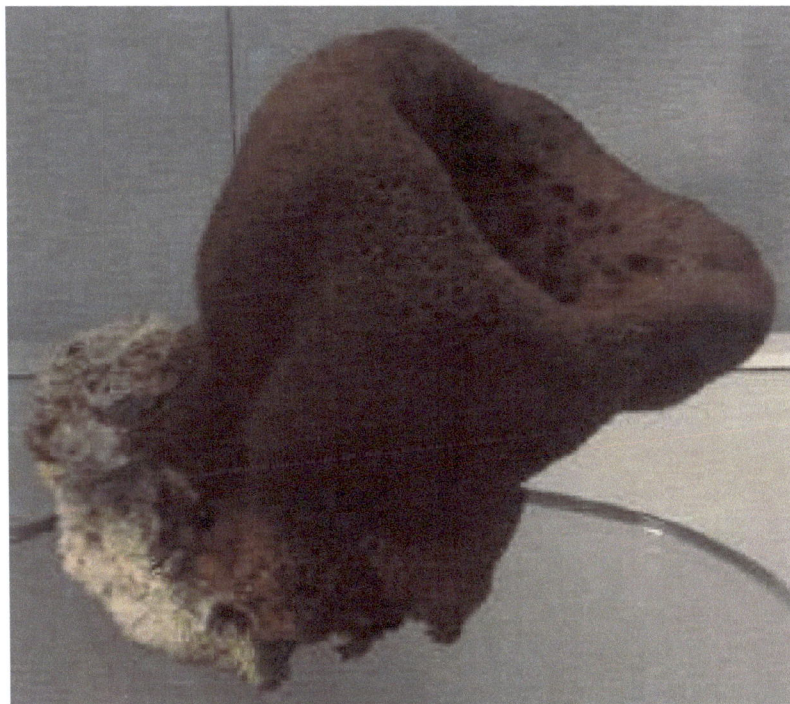

Fig. (42). *Hippospongia* sp. Image credit; Emőke Dénes; Creative Commons Attribution-Share Alike 4.0 International license. https://commons.wikimedia.org/wiki/File:Szi_-_Hippospongia_equina_1.jpg.

Antidiabetics and their mechanisms of action: Its compound, 9-Oxa-2-azabicyc-lo-[3,3,1]-nona-3,7-diene derivative (Fig. **43**), has been reported to show antidiabetic activity by inhibiting PTP1B with an IC_{50} value of 5.2 µM. Further, its 2-(Aminomethylene) hepta-3,5-dienedial moiety connected with the farnesyl group at C-7 (Fig. **44**) also showed similar activity with an IC_{50} value of 8.7 µM [24]. The isolation of seven sesterterpenoid compounds (1-7) from this species displayed PTP1B inhibitory activity [2], as detailed in Table **4**.

Table 4. PTP1B inhibitory activity of sesterterpenoid compounds of *Hippospongia lachne* [2].

Sesterterpenoid compounds (Figs. 45-50)	IC_{50} value (µM)
1	23.8
2	39.7
3	5.2
4	33
5	14
6	--
7	8.7

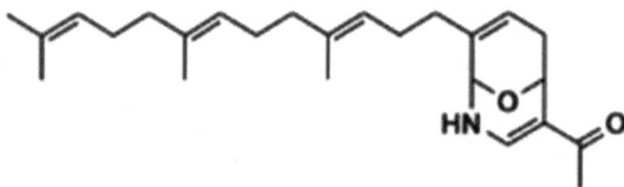

Fig. (43). 9-Oxa-2-azabicyclo-[3,3,1]-nona-3,7-diene derivative.

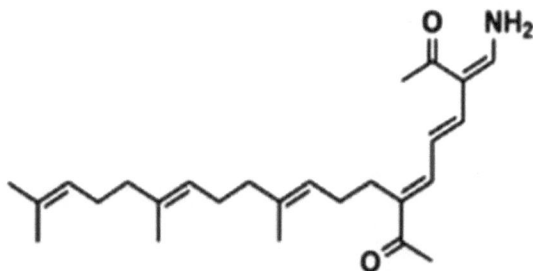

Fig. (44). 2-(Aminomethylene) hepta-3,5-dienedial moiety.

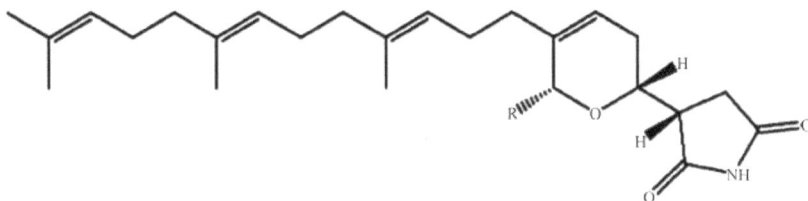

Fig. (45). Compound 1 (R=OH); Compound 2 (R= OCH3).

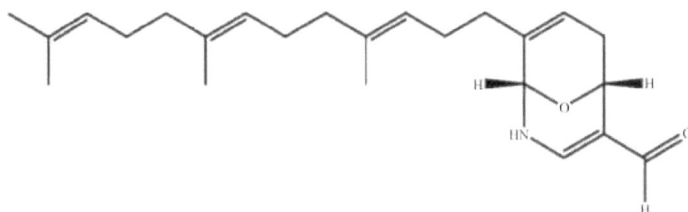

Fig. (46). Compound 3.

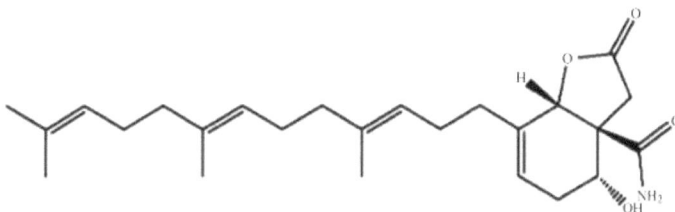

Fig. (47). Compound 4.

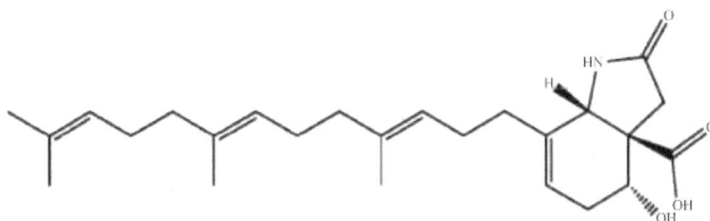

Fig. (48). Compound 5.

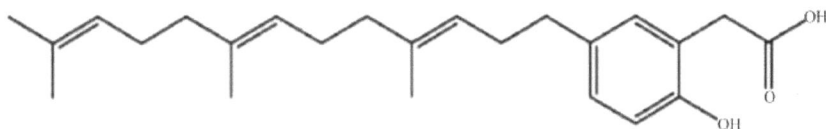

Fig. (49). Compound 6.

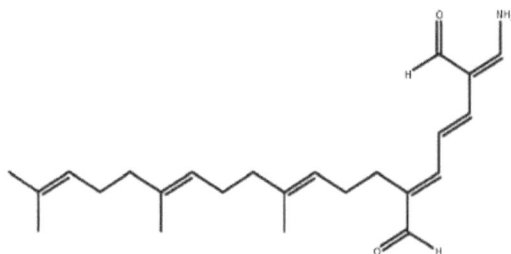

Fig. (50). Compound 7.

Hyattella sp.

Global distribution: It is a common species of Lakshadweep of India and Tahiti of French Polynesia.

Ecology: It is a cave-dwelling species.

Antidiabetics and their mechanisms of action: The pentacyclic scalarane sesterterpenes *viz.* hyattellactone A,B (Fig. **51**) and phyllofolactone F,G (Fig. **52**) have been derived from this species. Among these compounds, phyllofolactone F and hyattellactone A were found to inhibit PTP1B activity with IC_{50} values of 7.47 and 7.45 µM, respectively. On the other hand, the compounds hyattellactone B and phyllofolactone G were inactive [2].

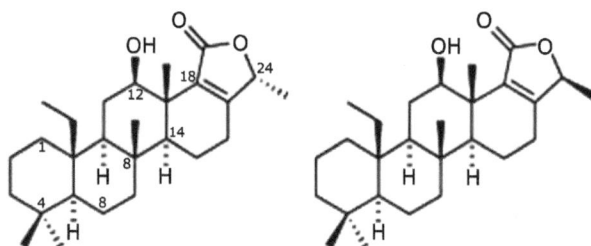

Fig. (51). Hyattellactone A,B.

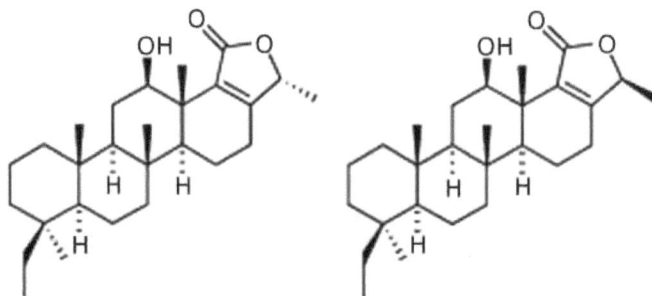

Fig. (52). Phyllofolactone F,G.

Hyrtios erectus

Global distribution: It is commonly found in the tropical Indo-Pacific: Palau and Hawaii (Fig. **53**).

Ecology: This sessile species has a depth range of 0-100 m.

Fig. (53). *Hyrtios erectus.* Image credit: Diego Delso ; license CC BY-SA ; https://commons.wikimedia.org/wiki/File:Esponja_%28Hyrtios_erectus%29,_mar_Rojo,_Egipto,_2023-04-15,_DD_85.jpg.

Antidiabetics and their mechanisms of action: Two scalarane-type sesterterpenes (Fig. **54**) derived from this species showed PTP1B inhibitory activity with IC_{50} values of 19.68 μM and 8.81 μM, respectively [2]. Its sesterterpenoid compound, hyrtiosal (Fig. **55**), displayed a dose-dependent non-competitive inhibitory activity against PTP1B with an IC_{50} value of 42 μM. Further, this compound also exhibited potent cellular effects on glucose transport, TGFbeta/Smad2 signaling, and PI3K/AKT activation [6].

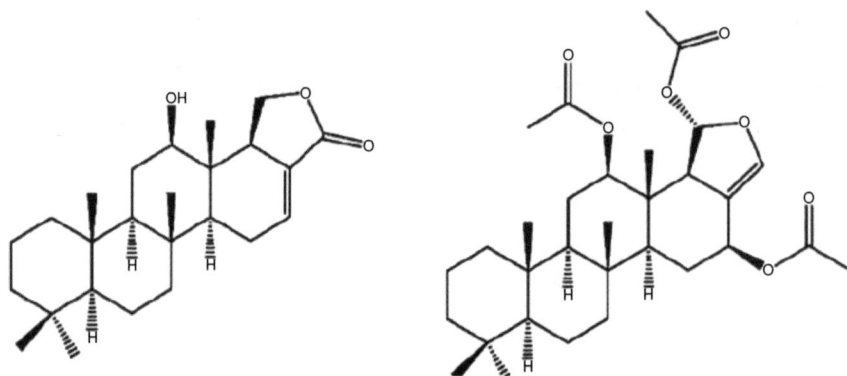

Fig. (54). Scalarane-type sesterterpenes.

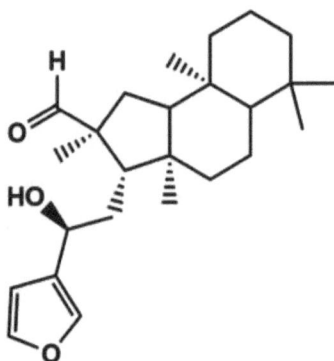

Fig. (55). Hyrtiosal.

Hyrtios sp.

Antidiabetics and their mechanisms of action: A meroterpenoid compound, nakijinol G (Fig. **56**), isolated from this species has shown promising PTP1B inhibitory activity with an IC_{50} value of 4.8 μM [2].

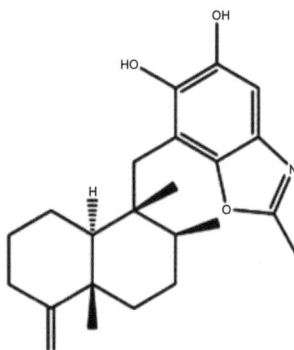

Fig. (56). Nakijinol G.

Ircinia dendroides

Global distribution: Tropical Western Central Pacific: Indonesia and Philippines. Ecology: This sessile species has been collected from trawl-exploitable bottoms (Fig. **57**).

Fig. (57). *Ircinia* sp. Image credit: Jodi Pirtle; public domain; https://commons.wikimedia.org/wiki/File: Ircinia_campana1.jpg.

Antidiabetics and their mechanisms of action: A sesquiterpene compound, palinurin (Fig. **58**), derived from this sponge species displayed antidiabetic activity with the inhibition of glycogen synthase kinase 3β (GSK-3β) [8, 25].

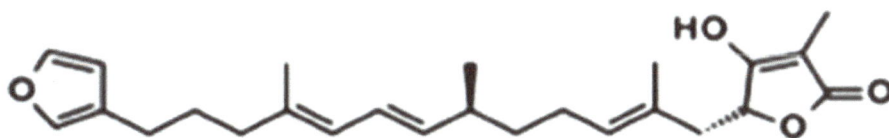

Fig. (58). Palinurin.

Ircinia sp.

Antidiabetics and their mechanisms of action: Two furanosesterterpenes *viz.* (7*E*, 12*E*, 20*Z*, 18*S*)-variabilin and (12*E*, 20*Z*, 18*S*)-8-hydroxyvariabilin (Fig. **59**), a terpenoid compound *viz.* furospongin-1 (Fig. **60**), and its semisynthetic derivative (Fig. **60**) of this sponge species have shown PTP1B inhibitory activity with IC_{50} values of 1.5, 7.1, 9.9, and 9.2 μM, respectively [2].

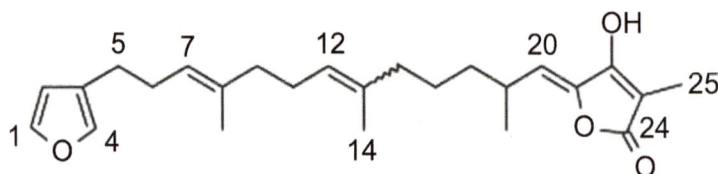

Fig. (59). (*7E*, *12E*, *20Z*, *18S*)-variabilin (84).

Fig. (60). Furospongin-1 (R=H). Derivative of Furospongin-1 (R=Ac).

Lamellodysidea herbacea

Global distribution: French Polynesia: Society islands.

Ecology: It is mostly seen on the inner reef barrier slope, in the lagoon, or on the pinnacles (Fig. **61**).

Fig. (61). *Lamellodysidea herbacea.* Image credit: Philippe Bourjon; Creative Commons Attribution-Share Alike 3.0 Unported license. https://commons.wikimedia.org/wiki/File:Eponge_%C3%A0_d%C3%A9 terminer_(1).jpg.

Antidiabetics and their mechanisms of action: The bioactive compounds isolated from this species have shown PTP1B inhibitory activity, as shown in Table **5**.

Table 5. PTP1B inhibitory activity of *Lamellodysidea herbacea* [8, 24].

Compound	IC$_{50}$ value (μM)
Polybromodiphenyl ether (Fig. **62**)	------
2-(3′,5′-Dibromo-2′-methoxyphenoxy)-3,5-dibromophenol (Fig. **63**)	0.9
2-(3′,5′-Dibromo-2′-methoxyphenoxy)-3,5-dibromophenol-methyl ether (Fig. **64**)	1.7
3,5-Dibromo-2-(3',5'-dibromo-2'-methoxyphenoxy)-1-methoxybenzene (Fig. **65**)	1.7
3,5-Dibromo-2-(3',5'-dibromo-2' -methoxyphenoxy)phenylethanoate (Fig. **65**)	0.6
3,5-Dibromo-2-(3',5'-dibromo-2' -methoxyphenoxy)phenylbutanoate (Fig. **66**)	0.7
3,5-Dibromo-2-(3',5'-dibromo-2' -methoxyphenoxy)phenylhexanoate (Fig. **66**)	0.7
3,5-Dibromo-2-(3',5'-dibromo-2' -methoxyphenoxy)phenyl benzoate (Fig. **67**)	1.0

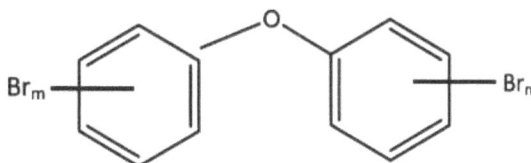

Fig. (62). Polybromodiphenyl ether.

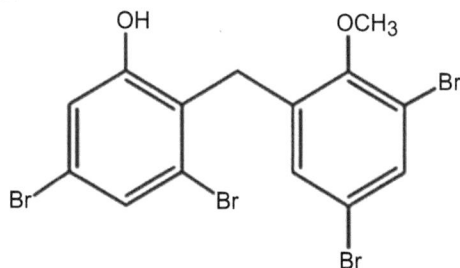

Fig. (63). 2-(3′,5′-Dibromo-2′-methoxyphenoxy)-3,5-dibromophenol.

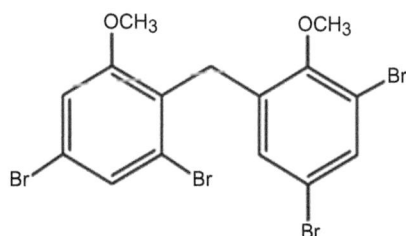

Fig. (64). 2-(3′,5′-Dibromo-2′-methoxyphenoxy)-3,5-dibromophenol-methyl ether.

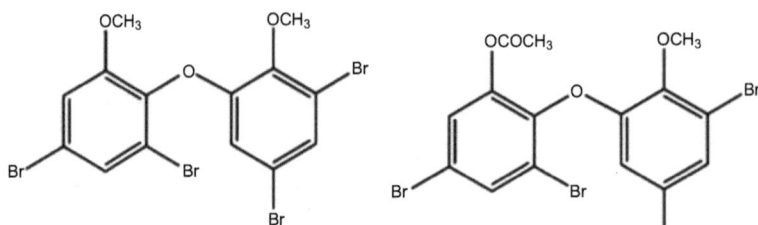

Fig. (65).   3,5-Dibromo-2-(3',5'-dibromo- dibromo-2' -2'-methoxyphenoxy)-1-methoxybenzene. 3,5-Dibromo-2-(3',5'- methoxyphenoxy)phenylethanoate.

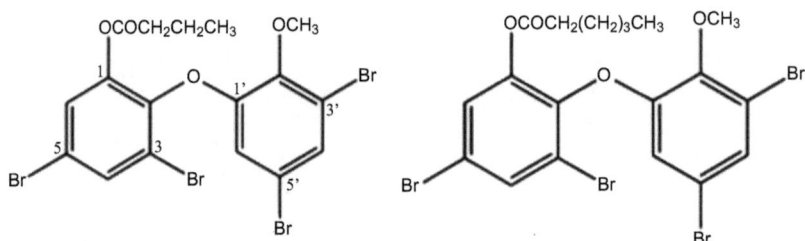

Fig. (66).   3,5-Dibromo-2-(3',5'-dibromo-2'-methoxyphenoxy)phenylbutanoate. 3,5-Dibromo-2-(3'-5'-dibromo-2' methoxyphenoxy)phenylhexanoate.

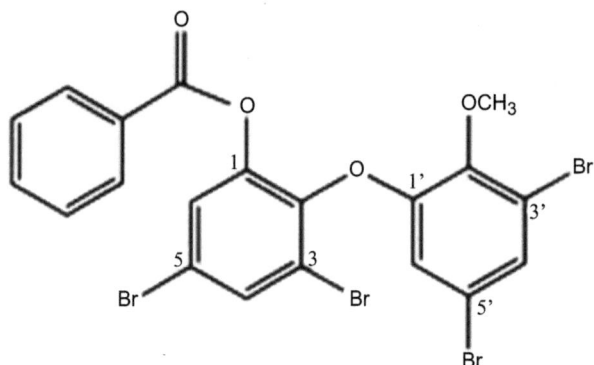

Fig. (67). 3,5-Dibromo-2-(3',5'-dibromo-2' –methoxyphenoxy)phenyl benzoate.

Mycale (Zygomycale) isochela

Global distribution: It has a wide distribution in Southwest Australia, the Northern Red Sea, and Egypt (Fig. **68**).

Ecology: This species is associated with mangrove and shallow seagrass bottoms in reef walls, lagoons, and bays at 10-25 m.

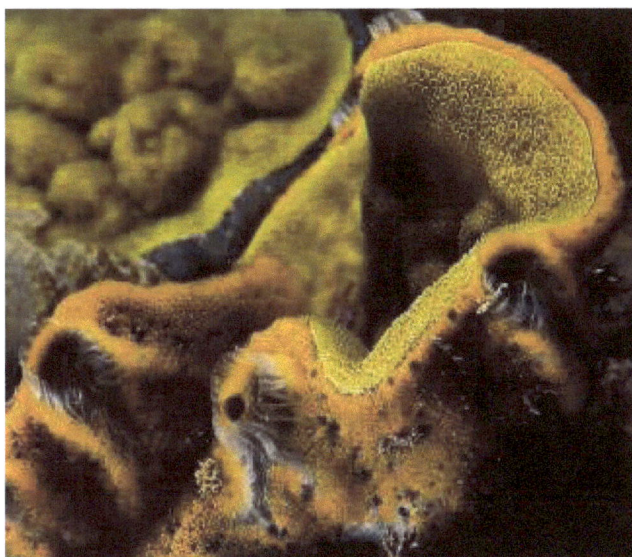

Fig. (68). *Mycale* sp. Image credit: Nhobgood Nick Hobgood; Creative Commons Attribution-Share Alike. 3.0 Unported, 2.5 Generic, 2.0 Generic, and 1.0 Generic license.; https://commons.wikimedia.org/wiki/File:Mycale_laevis_(Orange_Icing_Sponge)_around_hard_coral.jpg#m w-jump-to-license.

Antidiabetics and their mechanisms of action: A total of seven 5-alkylpyrrole- 2-carboxaldehyde derivatives, *viz.* mycalenitriles (Fig. **69**), have been isolated from this species. Except compound 98, all other compounds have shown antidiabetic activity by inhibiting PTP1B, and the IC50 values recorded for compounds were 8.6, 10.0, and 3.1 μM (1-3; 117-119), respectively, and 26.2, 28.2, and 12.5 μM (5-7; 121-123), respectively [2].

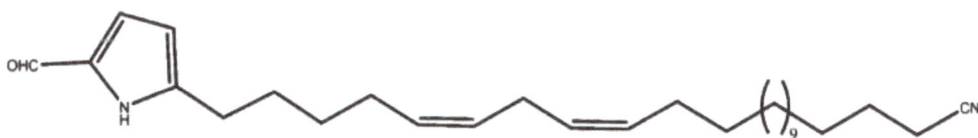

Fig. (69). 5-alkylpyrrole-2-carboxaldehyde derivatives (mycalenitriles).

Mycale sp.

Antidiabetics and their mechanisms of action: The compound, 5α,8α-Epidioxycholest-6,22-dien-3β-ol derived from this species, has shown antidiabetic activity [24].

5α,8α-Epidioxycholest-6,22-dien-3β-ol

Niphates sp.

Antidiabetics and their mechanisms of action: The extracted metabolites of this sponge species have shown both alpha-glucosidase and amylase inhibitory activities, and the IC50 values recorded for the different isolates are given in Table **6**, (Fig. **70**).

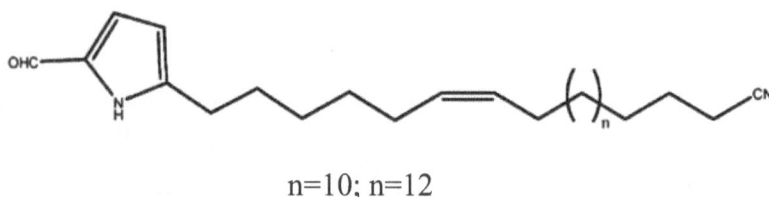

n=10; n=12

Fig. (70). *Niphates* sp. Image credit: Twilight Zone Expedition Team 2007, NOAA-OE; public domain ; https://commons.wikimedia.org/wiki/File:Sponges_in_Caribbean_Sea,_Cayman_Islands.jpg.

Table 6. Alpha-glucosidase and amylase inhibitory activities of the extracts of *Niphates* sp.[96].

Isolate	Alpha-Glucosidase Inhibitory Activity (IC50 ug/ml)	Amylase Inhibitory Activity (IC50 ug/ml)
QH 26	177.6	9.0
QH 30	453.8	25.0
QH 36	153.5	15.6
QH 37	441.0	19.3
QN 45	341.8	21.5
QN 63	495.4	201.9

Oceanapia triangulata (= Pellina triangulata)

Global distribution: West Atlantic, West Pacific, and Indian Ocean. Ecology: It is a sessile species.

Antidiabetics and their mechanisms of action: A C_{31} polyacetylenic acid *viz.* corticatic acid (Fig. **71**) derived from this species has shown α-glucosidase inhibitory activity with an IC_{50} value of 0.16 μg/mL.

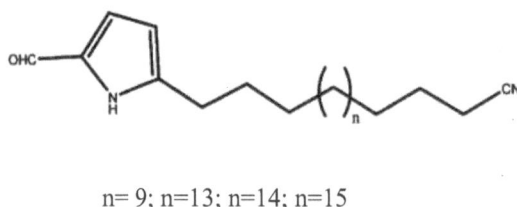

n= 9; n=13; n=14; n=15

Fig. (71). Corticatic acid.

Penares schulzei

Global distribution: It is a common species of tropical Pacific surrounding New Caledonia.

Ecology: It is a benthic species.

Antidiabetics and their mechanisms of action: Three isoquinoline-derived alkaloids *viz.* schulzeines A–C (Figs. **72-74**) derived from this sponge species demonstrated α-glucosidase inhibitory effects with IC_{50} values ranging from 48 to 170 nM [2].

Fig. (72). Schulzeine A.

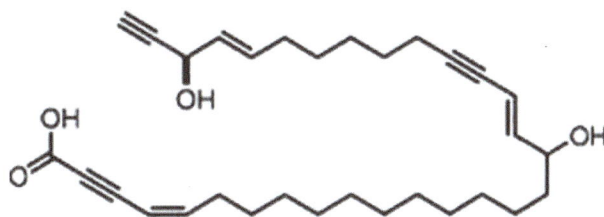

Fig. (73). Schulzeine B.

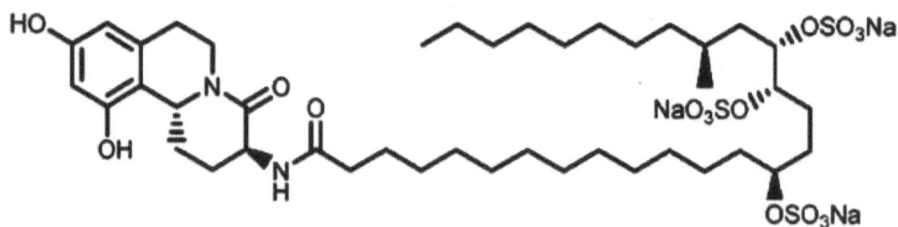

Fig. (74). Schulzeine C.

Penares sp.

Antidiabetics and their mechanisms of action: The compounds penasulfate A, A1 and A2 (Figs. **75-78**), derived from this sponge species, displayed α-glucosidase inhibitory effects with IC_{50} values of 3.5, 1.2, and 1.5 mg/mL, respectively [2].

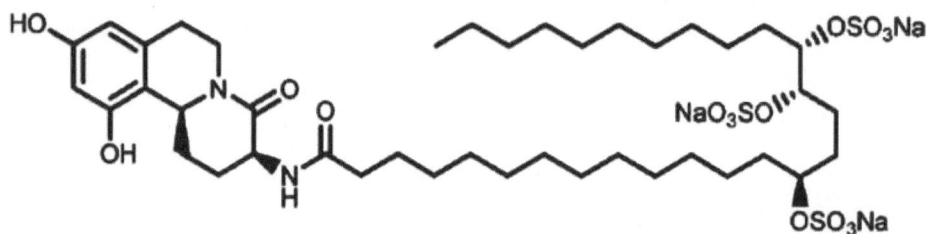

Fig. (75). *Plakortis* sp.Image credit: Philippe Bourjon; Creative Commons Attribution-Share Alike 4.0 International license. https://commons.wikimedia.org/wiki/File:Plakortis_sp._2_%C3%A0_confirmer.jpg.

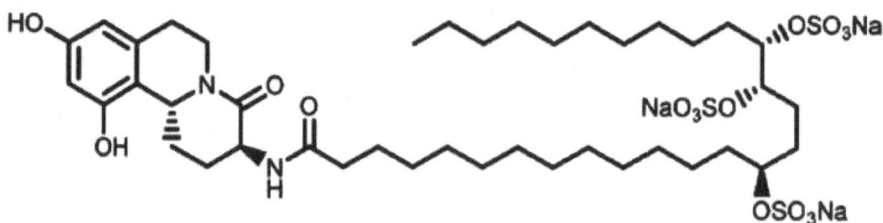

Fig. (76). Penasulfate A.

Fig. (77). Penasulfate A1.

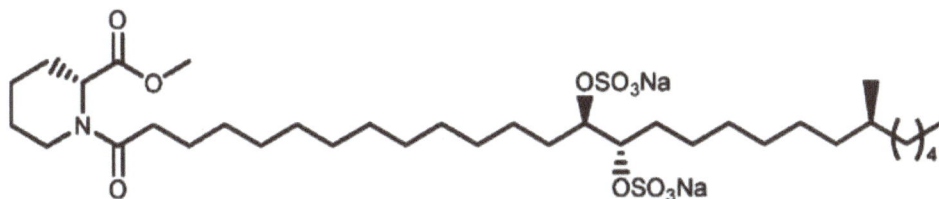

Fig. (78). Penasulfate A2.

Petrosia (Strongylophora) strongylata (= Strongylophora strongylata)

Global distribution: Mediterranean Sea and Eastern Atlantic.

Ecology: It resides on reef slopes, overhangs, underside of rocks, and in caves at depths of 5 - 70 m.

Antidiabetics and their mechanisms of action: The diterpenoid strongylophorines derived from this species have shown PTP1B inhibitory activities, as shown in Table **7**.

Table 7. PTP1B inhibitory activities of strongylophorines of *Petrosia (Strongylophora) strongylata* [2].

Strongylophorine	PTP1B inhibitory activity (IC$_{50}$, µM).
26-*O*-ethylstrongylophorine-14 (Fig. **79**)	8.7
26-*O*-methylstrongylophorine-16 (Fig. **80**)	8.5

(Table 7) cont.....

Strongylophorine	PTP1B inhibitory activity (IC$_{50}$, μM).
Strongylophorine-2 (Fig. **81**)	24.4
Strongylophorine-3(Fig. **82**)	9.0
Strongylophorine-8 (Fig. **83**)	21.2
Strongylophorine-15 (Fig. **84**)	11.9
Strongylophorine-17 (Fig. **85**)	14.8

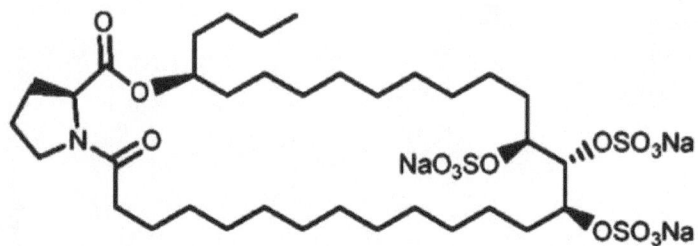

Fig. (79). 26-*O*-ethylstrongylophorine-14.

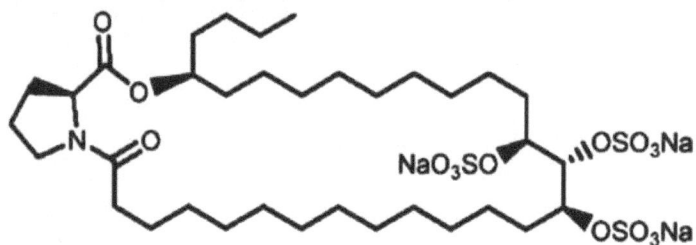

Fig. (80). 26-*O*-methylstrongylophorine-16.

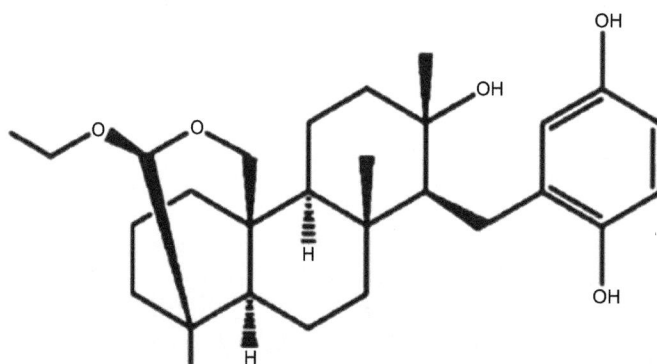

Fig. (81). Strongylophorine-2.

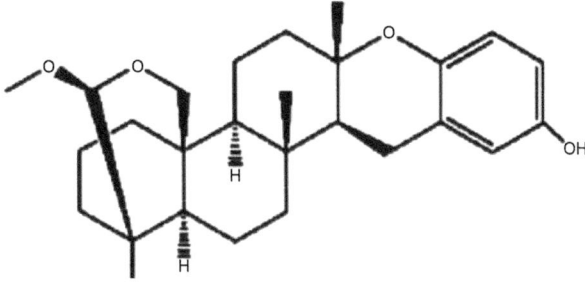

Fig. (82). Strongylophorine-3.

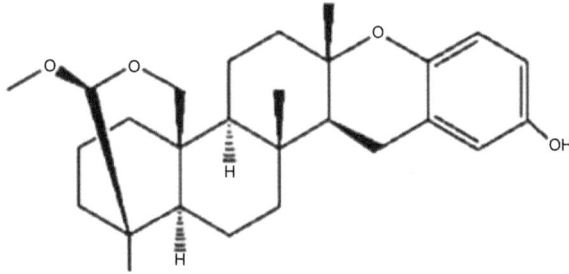

Fig. (83). Strongylophorine-8.

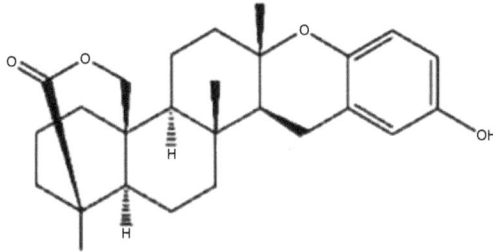

Fig. (84). Strongylophorine-15.

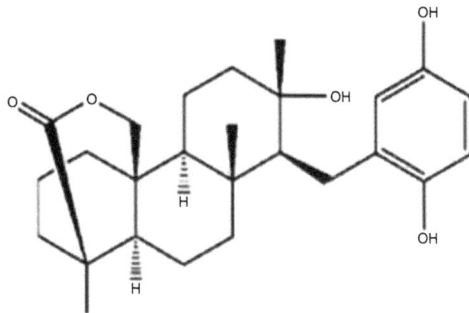

Fig. (85). Strongylophorine-17.

Petrosia sp.

Antidiabetics and their mechanisms of action: A tetramic acid derivative *viz.* melophlin C (Fig. **86**) of this sponge species has shown PTP1B inhibitory activity with an IC_{50} value of 14.6 µM [2].

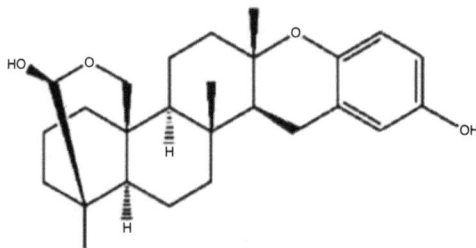

Fig. (86). Melophlin C.

Plakortis simplex

Global distribution: It is a temperate species of Central Pacific and NE Atlantic. Ecology: This sessile and boring species is observed only during winters.

Antidiabetics and their mechanisms of action: A polyketide compound, woodylide C (Fig. **87**), isolated from this sponge species has shown inhibitory activity against PTP1B with an IC_{50} of 4.7 µg/mL [2].

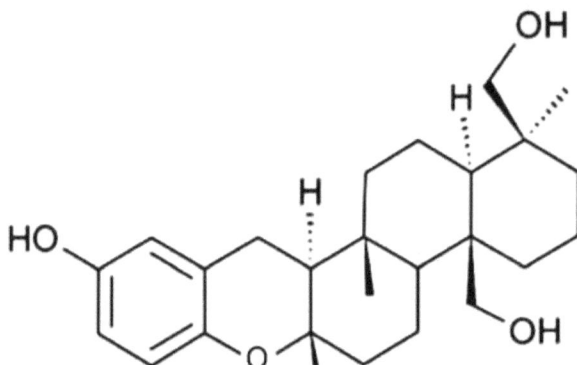

Fig. (87). Woodylide C.

Siphonodictyon coralliphagum (= Aka coralliphaga)

Global distribution: It is found throughout the Caribbean.

Ecology: It is a shallow water, coral reef boring sponge occurring at depths of 10-70 m (Fig. **88**).

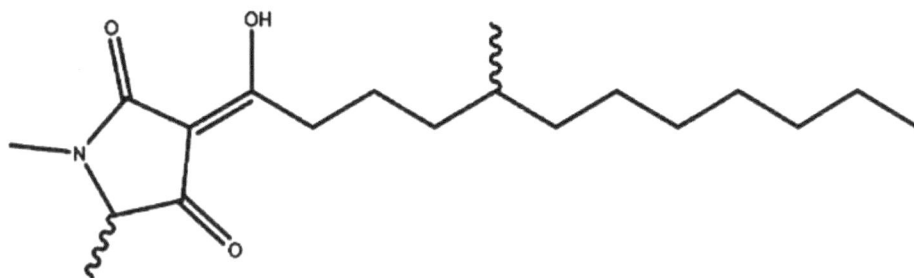

Fig. (88). *Siphonodictyon coralliphagum* Image credit: iNaturalist user: thibaudaronson; Creative Commons Attribution-Share Alike 4.0 International license. https://commons.wikimedia.org/wiki/File:Siphonodictyon_coralliphagum.jpg

Antidiabetics and their mechanisms of action: A total of 41 strains of bacteria belonging to the phyla Firmicutes (23), Actinobacteria (9), Proteobacteria (7), and Bacteroidetes (1) have been found associated with this sponge species [25, 28]. All these strains have been reported to possess β-Glucosidase inhibition activity, as shown in Table **8**.

Table 8. β-Glucosidase inhibition activity of bacterial strains associated with *Siphonodictyon coralliphagum* [28].

Bacterial Strain	β-Glucosidase Inhibitory Activity (mm)
Stenotrophomonas rhizophila SD1-25	98.5
Bacillus flexus SD2-1	100
Bacillus oceanisediminis SD2-2(1)	99.5
Bacillus siamensis SD2-2(2)	100
Bacillus flexus SD2-3(2)	100
Bacillus sp. SD2-5	100
Arthrobacter koreensis SD2-6(1)	100
Bacillus stratosphericus SD2-7(2)	100
Exiguobacterium sp. SD2-15(1)	91.0
Bacillus oceanisediminis SD2-17	99.1
Stenotrophomonas rhizophila SD1-1	99.8
Arthrobacter koreensis SD1-3	100
Advenella kashmirensis SD1-6(1)	99.0
Microbacterium oleivorans SD1-8	99.7
Arthrobacter koreensis SD1-13	100
Planomicrobium okeanokoites SD1-14(1)	99.8

(Table 8) cont.....

Bacterial Strain	β-Glucosidase Inhibitory Activity (mm)
Dietzia maris SD1-17	97.3
Chryseomicrobium sp SD1-18	96.0
Sphingobacterium sp SD1-20(1)	90.0
Exiguobacterium marinum SD1-23	97.3
Bacillus aryabhattai GPB13	100
Bacillus aryabhattai GPB20	99.6
Staphylococcus gallinarum GPB21	100
Halomonas sulfidaeris SP2B3	97.8
Bacillus tequilensis SP2B5	99.4
Bacillus sp SP2B6	99.5
Leucobacter chromiiresistens SP2B9	99.7
Planococcus rifitoensis SP2B11	98.8
Bacillus stratosphericus SP2B12	99.5
Bacillus amyloliquefaciens subsp. *amyloliquefaciens* SP2B20	100
Streptomyces rangoonensis SP2A6	97.6
Microbacterium esteraromaticum SD2-18	99.8
Bacillus methylotrophicus SD2-20	98.7
Bacillus subtilis subsp. *inaquosorum* SD2-22	99.8
Psychrobacter maritimus SD2-24	99.7
Vibrio communis GDN4	99.6
Bacillus sp. GDB16	95.0
Streptomyces coelicoflavus GDA11	99.6
Staphylococcus gallinarum GPB8	99.7
Bacillus subtilis subsp. *spizizenii* GPB9	99.7
Pseudochrobactrum sp. GPB10	96.0

Spongia sp.

Global distribution: It is endemic to the Nearctic Region and is only found in Mexico, Costa Rica, and Florida (Fig. **89**).

Ecology: This shallow water species (1 - 10 m below the surface) is seen up to a depth of 100 m.

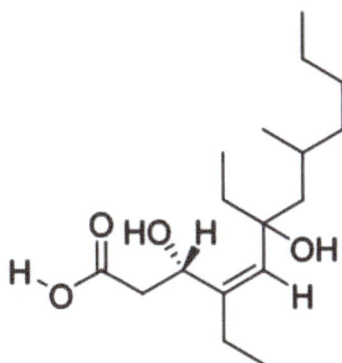

Fig. (89). *Spongia* sp. Image credit: H. Zell; Creative Commons Attribution-Share Alike 3.0 Unported license. https://commons.wikimedia.org/wiki/File:Spongia_officinalis_001.JPG.

Antidiabetics and their mechanisms of action: Two furanosesterterpenes *viz.* furospongin-1 and its semisynthetic derivative (Fig. **90**) of this sponge species have been reported to show PTP1B inhibitory activity with IC_{50} values of 9.9 and 9.2 µM, respectively [2].

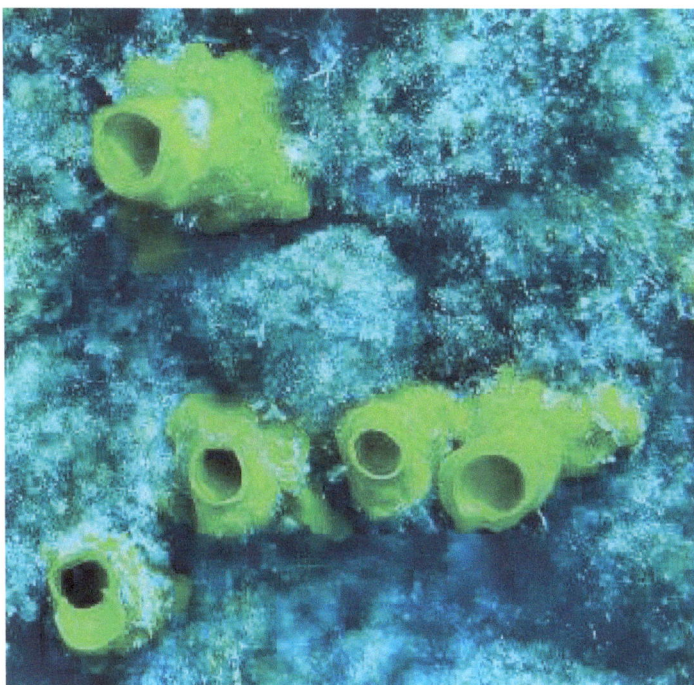

Fig. (90). Furospongin-1 (R=H); Semisynthetic derivative (R=Ac).

Stelletta sp.

Global distribution: NE Atlantic.

Ecology: It is commonly seen in caves or beneath overhangs. It is also found in tidal streams of the open coast and sea loughs (Fig. **91**).

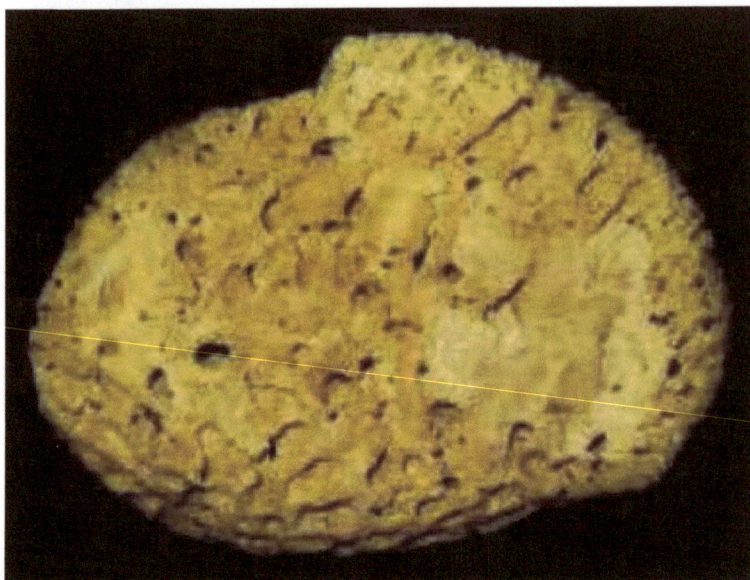

Fig. (91). *Stelletta* sp. Image credit: Peter Southwood ; Creative Commons Attribution-Share Alike 4.0 International license.; https://commons.wikimedia.org/wiki/File:Stelletta_agulhana_14169990.jpg.

Antidiabetics and their mechanisms of action: An isomalabaricane triterpene *viz.* stellettin N and its analogs, stellettin H, rhabdastrellic acid A, stellettin G, stellettin D, and 22,23-dihydrostellettin (Figs. **92-97**), have been derived from this sponge species. Among these compounds, only stellettin G has been reported to yield PTP1B inhibitory activity with an IC_{50} value of 4.1 μM [6, 24].

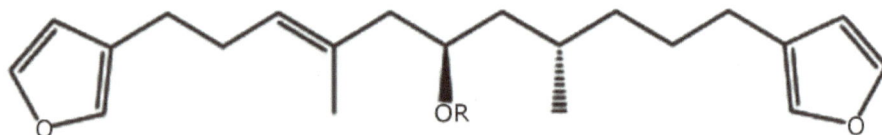

Fig. (92). Stellettin N.

Fig. (93). Stellettin H.

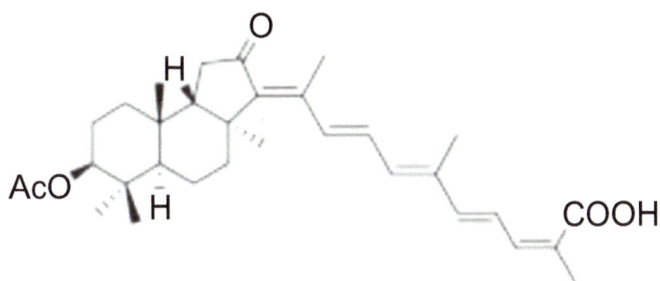

Fig. (94). Rhabdastrellic acid A.

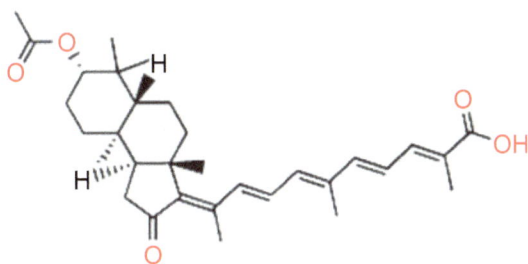

Fig. (95). Stellettin G.

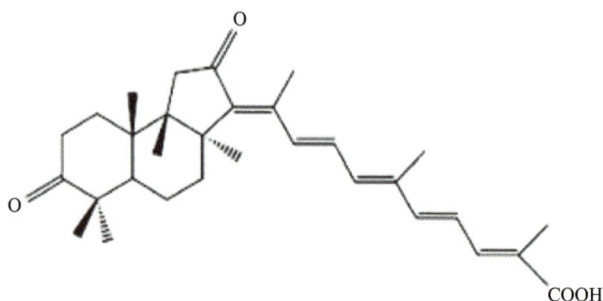

Fig. (96). Stellettin D.

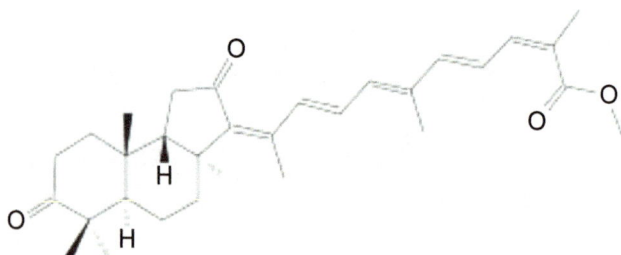

Fig. (97). 22,23-dihydrostellettin.

Xetospongia muta

Global distribution: It is a common species of Caribbean coral reefs. Ecology: This sessile species has a depth range of 10-120 m (Fig. **98**).

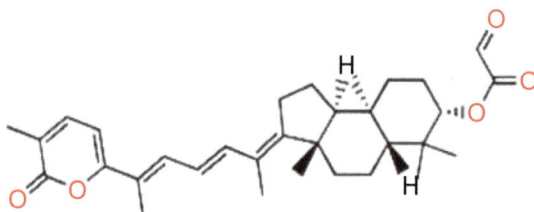

Fig. (98). *Xsetospongia muta.* Image credit: Hickerson/FGBNMS; public domain; https://commons.wikimedia.org/wiki/File:Xestospongia_muta_1.jpg.

Antidiabetics and their mechanisms of action: The aqueous extracts of this sponge species showed dipeptidyl peptidase IV inhibitory effects [24].

Xestospongia testudinaria

Global distribution: It is a common species of Indo-Pacific. Areas of its dominance include Indonesia, the Caribbean Sea, the Bahamas, Bermuda, the Gulf of Mexico, and Florida (Fig. **99**).

Ecology: It is commonly seen at depths of 10-120m.

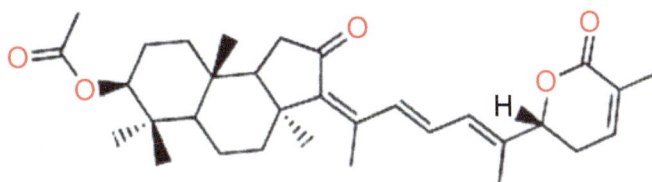

Fig. (99). *Xestospongia testudinaria* . Image credit: Bernard DUPONT from FRANCE; Creative Commons Attribution-Share Alike 2.0 Generic license.; https://commons.wikimedia.org/wiki/File:Barrel_Sponge_%28Xestospongia_testudinaria%29_%288500727224%29.jpg.

Antidiabetics and their mechanisms of action: An unidentified sterol compound (Fig. **100**) and a brominated lipid compound (Fig. **101**) isolated from this species exhibited PTP1B inhibitory effects with IC_{50} values of 4.3 and 5.3 μM, respectively [2]. Among its 24-Hydroperoxy-24-Vinylcholesterol and 29-Hydroperoxystigmasta-5,24(28)-dien-3-ol (Figs. **102, 103**), only the latter has shown similar activity with an IC_{50} value of 5.8 μg/mL [24].

Fig. (100). An unidentified sterol compound.

Fig. (101). An unidentified brominated lipid compound.

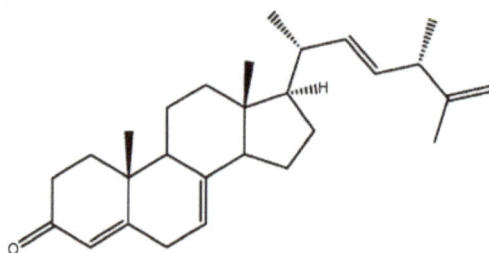

Fig. (102). 24-Hydroperoxy-24-Vinylcholesterol.

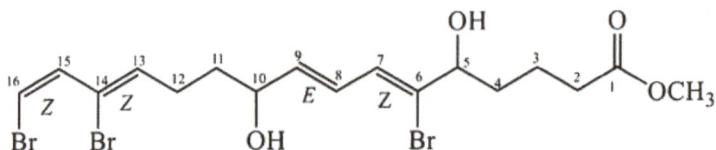

Fig. (103). 29-Hydroperoxystigmasta-5,24(28)-dien-3-ol.

Cnidarians

Bartholomea annulata

Global distribution: It is a common species in the Gulf of Mexico and the Caribbean (Fig. **104**).

Ecology: This epibenthic and solitary species is normally found on coral reefs and rocky bottoms in subtidal areas. It is also seen in mangroves.

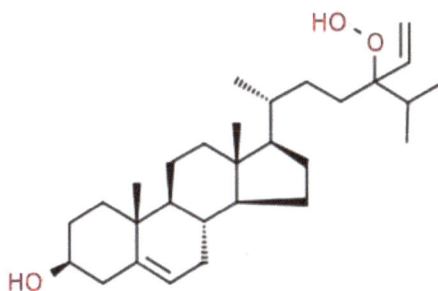

Fig. (104). *Bartholomea annulata* Fernando Herranz Martín; Creative Commons CC0 License; https://commons.wikimedia.org/wiki/File:Bartholomea_annulata.jpg.

Antidiabetics and their mechanisms of action: The aqueous extracts of this species exhibited antidiabetic effects by inhibiting the enzyme Dipeptidyl peptidase IV [8].

Bunodosoma granuliferum

Global distribution: It is widely distributed in West Atlantic. Ecology: It is seen in all marine habitats and at all depths (Fig. **105**).

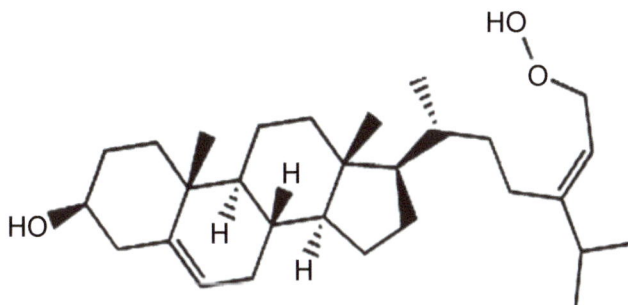

Fig. (105). *Bunodosoma granulifera* . Ricardo González-Muñoz, *et al*., Creative Commons Attribution 3.0 Unported license.; https://commons.wikimedia.org/wiki/File:Bunodosoma_granuliferum.jpg.

Antidiabetics and their mechanisms of action: The water extracts of this species have shown antidiabetic activity by inhibiting the enzyme Dipeptidyl peptidase IV [8].

Clavularia viridis

Global distribution: It is native to the tropical Indo-Pacific.

Ecology: It is known to occur at a depth range of 0-20m, and it is known to colonize other species of coral (Fig. **106**).

Fig. (106). *Clavularia viridis.(*Max Rühle; Creative Commons Attribution 2.0 Generic license.; https://commons.wikimedia.org/wiki/File:Clavularia_virdis_-_R%C3%B6hrenkoralle.jpg.

Antidiabetics and their mechanisms of action: The dolabellanes diterpene, clavurol E (Fig. **107**), derived from this species has shown PTP1B inhibitory activity with an IC50 value of 14.5 µg/mL [2],

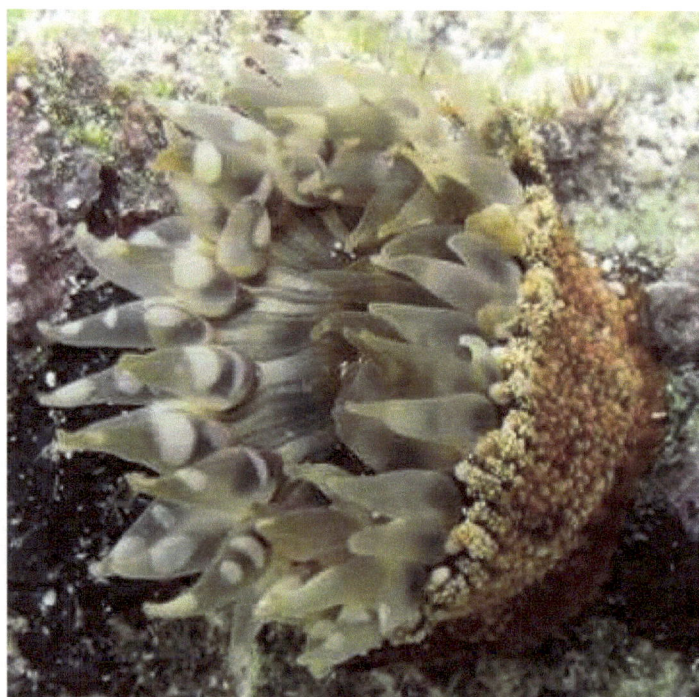

Fig. (107). Clavurol E.

Dichotella gemmacea

Global distribution: Tropical Indo-Pacific.

Ecology: It is known to prefer deeper waters and inhabits normally in coral reefs with rocky substrates.

Antidiabetics and their mechanisms of action: Its steroid derivative, 5α,8α-epidioxy-22E-ergosta-6,22-dien-3β-ol (ergosterol peroxide) (Fig. **108**), and sterol, 3β-Hydroxycholest-5-en-25-acetoxy-19-oate (Fig. **109**), derived from this species, have shown no obvious activity against PTP1B (IC 50 values 4100 mg/mL) [24].

Fig. (108). 5α,8α-Epidioxy-ergosta-6,22-dien-3β-ol (Ergosterol peroxide).

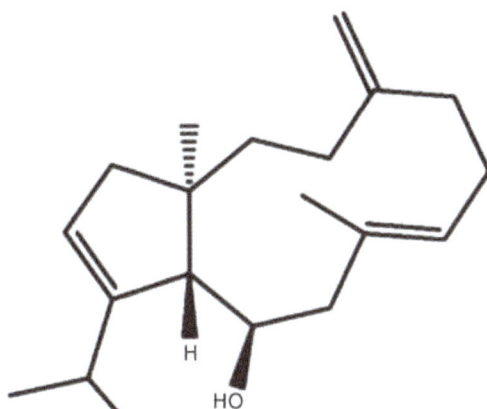

Fig. (109). 3β-Hydroxycholest-5-en-25-acetoxy-19-oate.

Eunicea sp.

Global distribution: Throughout the Caribbean province.

Ecology: It dwells in shallow to mid-depth areas with strong wave surges (Fig. **110**).

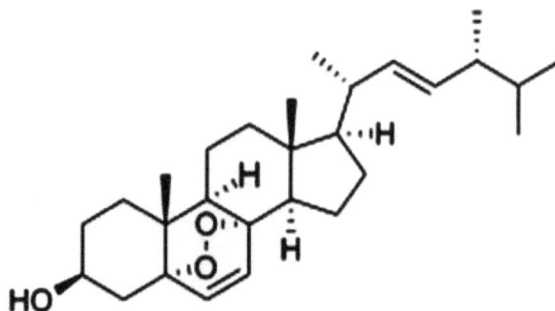

Fig. (110). *Eunicea* sp. Quintín Muñoz ; Creative Commons CC0 License ; https://commons.wikimedia.org/wiki/File:Eunicea_mammosa1.jpg.

Antidiabetics and their mechanisms of action: Two new cembranoids *viz.*

14-deoxycrassin and pseudoplexauric acid methyl ester (Figs. **111, 112**) derived from the extracts of this species have been reported to activate the proliferation of pancreatic beta-cells in order to counteract hyperglycemia [27].

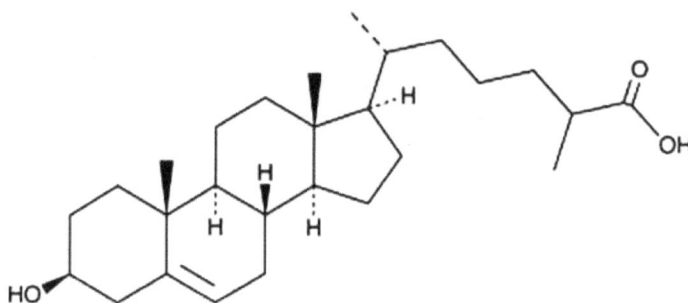

Fig. (111). 14-deoxycrassin.

Lobophytum pauciflorum

Global distribution: Andaman and Lakshadweep Islands and Indo-West Pacific. Ecology: It is found mostly in reef habitats of shallow waters (Fig. **113**).

Antidiabetics and their mechanisms of action: The crude extracts of this species demonstrated antidiabetic effects with the reduction of plasma glucose levels in experimental rats [8].

Radianthus magnifica (= Heteractis magnifica)

Global distribution: Tropical Indo-Pacific Ocean, from the Red Sea to Samoa.
Ecology: It is found attached to rocks in shallow reefs (Fig. **114**).

Fig. (112). Pseudoplexauric acid methyl ester.

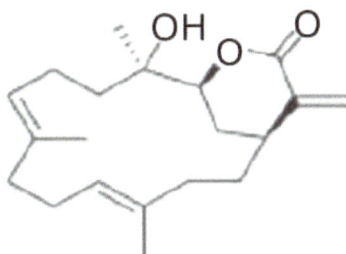

Fig. (113). *Lobophytum pauciflorum* David Witherall; Creative Commons Attribution 3.0 Unported license.;
https://commons.wikimedia.org/wiki/File:Lobophytum_pauciflorum.jpeg.

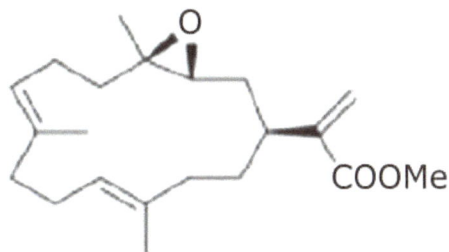

Fig. (114). *Radianthus magnifica* Nhobgood Nick Hobgood; Creative Commons Attribution-Share Alike 3.0
Unported license.
https://en.wikipedia.org/wiki/File:Amphiprion_perideraion_(Pink_anemonefish)_in_Heteractis_magnifica_(
Magnificent_sea_anemone).jpg.

Antidiabetics and their mechanisms of action: The mucus of this species showed the presence of a peptide *viz*. magnificamide, which possesses antidiabetic activity by inhibiting the enzyme α-amylase. Further, this compound was found to act against human salivary amylase and porcine pancreatic amylase with Ki (inhibitory constant) values of 7.7 and 0.17nM, respectively [2].

Sarcophyton glaucum

Global distribution: It is a common species of south equatorial East African reefs.

Ecology: It is typically found in intertidal, subtidal, and near-shore reef flat habitats (Fig. **115**).

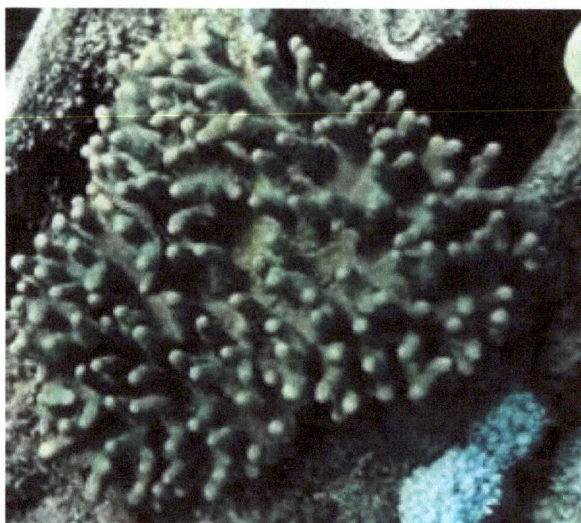

Fig. (115). *Sarcophyton glaucum*. Karelj; public domain.; https://commons.wikimedia.org/wiki/File:Sarcopython_glaucum_1.jpg.

Antidiabetics and their mechanisms of action: The crude extracts of this species demonstrated antidiabetic effects with the reduction of plasma glucose levels in experimental rats [8].

Sarcophyton subviride (-associated fungus *Aspergillus* terreus)

Global distribution: It is widely distributed across the globe, especially in the Indo-Pacific region.

Ecology: It is a species of shallow coastal waters.

Antidiabetics and their mechanisms of action: The dimeric sesquiterpene lactone, versicolactone G (Fig. **116**), isolated from the fungus Aspergillus terreus associated with this species, has shown antidiabetic activity by potently inhibiting α-glucosidase [97].

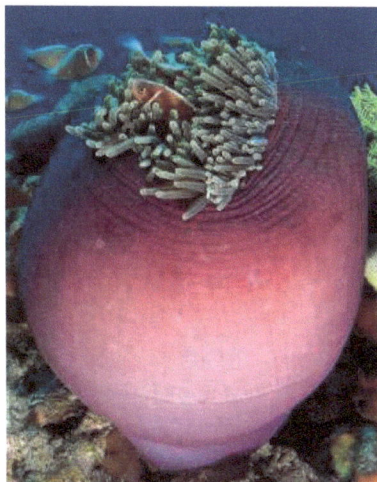

Fig. (116). Versicolactone G.

Sarcophyton trocheliophorum

Global distribution: Indo-pacific: South China Sea and off the coasts of Saudi Arabia and Egypt (Fig. **117**).

Ecology: It is commonly found in coral reef areas.

Fig. (117). *Sarcophyton trocheliophorum*. Philippe Bourjon; Creative Commons Attribution-Share Alike 3.0 Unported license.; https://commons.wikimedia.org/wiki/File:Sarcophyton_trocheliophorum.jpg.

Antidiabetics and their mechanisms of action: The compounds with a diterpenoid moiety and a polyhydroxylated steroid, 7α-hydroxy-crassarosterol A, derived from this species, have shown antidiabetic activity by serving as PTP1B inhibitors [2]. The IC50 values recorded for these compounds are shown in Table **9**.

Table 9. PTP1B inhibitory activity of Sarcophyton trocheliophorum [2].

Compound	PTP1B Inhibitory Activity (IC50, μM)
Sacrophytonolide N (Fig. **118**)	5.95
Cembrene-C (Fig. **118**)	26.6
Ketoemblide (Fig. **119**)	27.2
4Z,12Z,14E-sarcophytolide (Fig. **120**)	15.4
Sarcrassin E (Fig. **120**)	6.33
Sartrolide H (Fig. **121**)	19.9
Sarcotroate B (Fig. **122**)	6.97
Sarsolilide A (Fig. **122**)	6.8
Sarsolide B (Fig. **122**)	27.1
Secodihydrosarsolenone (Fig. **122**)	13.7
(E,E,E)-1-isopropenyl-4,8,12-trimethylcyclotetradeca-3,7,11-triene (Fig. **123**)	22.19
Sarcophytonolide I (Fig. **124**)	11.26
7α-hydroxy-crassarosterol A (Fig. **125**)	33.05

Its diterpenoids *viz.* methyl sarcotroate A and B (Fig. **126**) displayed PTP1B inhibitory activity with an IC 50 value of 7.0 μM [24].

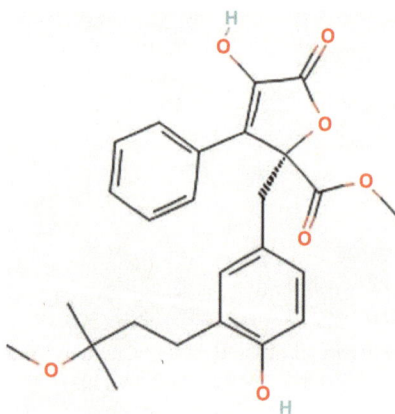

Fig. (118). Sacrophytonolide N.

Fig. (119). Ketoemblide.

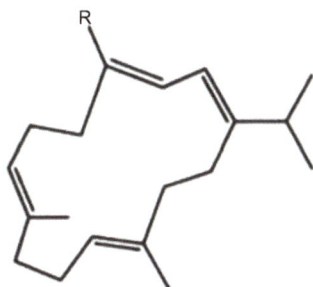

Sacrophytonolide N (R= COOCH3)

Cembrene-C (R= CH3)

Fig. (120). Sarcrassin E.

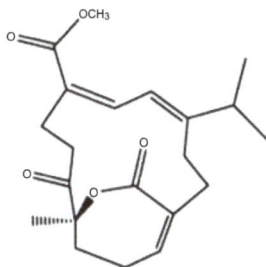

Fig. (121). Sartrolide H.

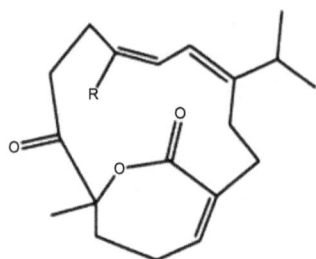

4Z,12Z,14E-sarcophytolide (R=CH3).

Sarcrassin E (R=COOCH3)

Fig. (122). Sarcotroate B, Sarsolilide A, Sarsolilide B, Secodihydrosarsolenone.

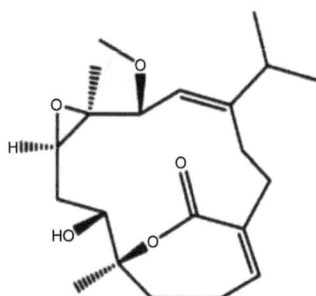

Fig. (123). (E,E,E)-1-isopropenyl-4,8,12-trimethylcyclotetradeca-3,7,11-triene.

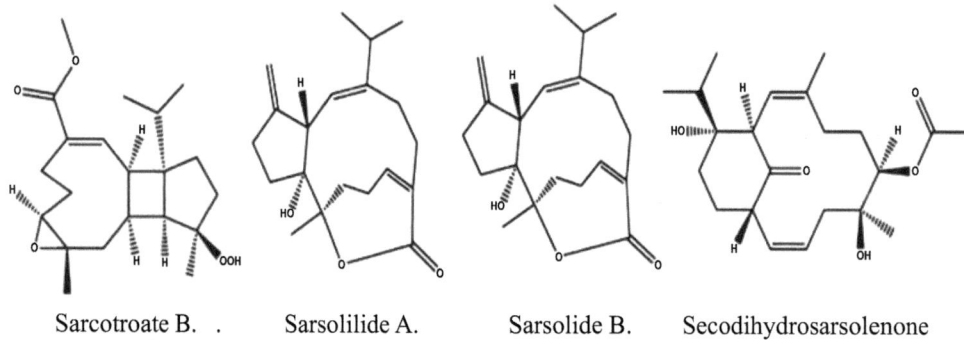

Sarcotroate B.   .   Sarsolilide A.   Sarsolide B.   Secodihydrosarsolenone

Fig. (124). Sarcophytonolide I.

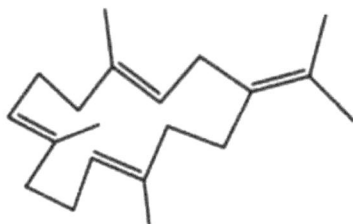

Fig. (125). 7α-hydroxy-crassarosterol A.

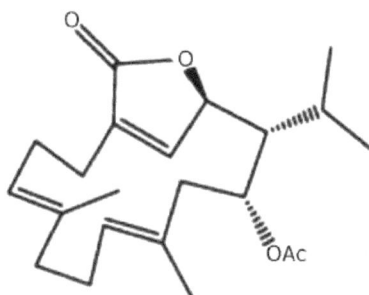

Fig. (126). Methyl sarcotroate A, Methyl sarcotroate B.

Sclerophytum depressum (= Sinularia depressa)

Global distribution: It has a global distribution.

Ecology: It is believed to occur in both marine and freshwater environments (Fig. **127**).

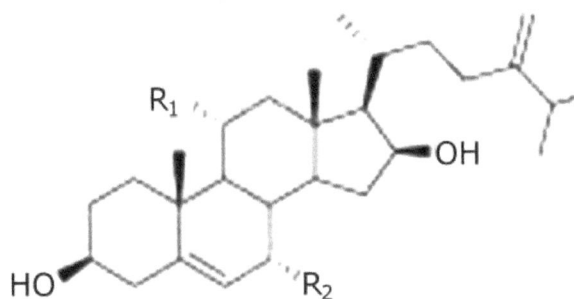

R1= CH3; R2= OH

Fig. (127). *Sclerophytum depressum.* Marie HENNION; Creative Commons Attribution 4.0 International license.; https://commons.wikimedia.org/wiki/File:Sinularia_depressa_%28MNHN-IK-2015-2175%29.jpeg.

Antidiabetics and their mechanisms of action: Three unidentified steroids (1-3) (Figs. **128, 129**), including a monoacetylated derivative of this species, have shown potent PTP1B inhibitory activity with IC50 values of 15.3, 19.5, and 22.7 μM, respectively [2].

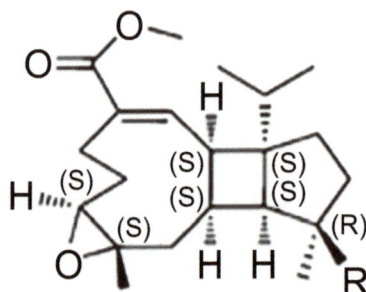

Methyl sarcotroate A (R=OH)

Methyl sarcotroate B (R= OOH)

Fig. (128). Sterol 1. Sterol 2 (R1=OH; R2=OH).

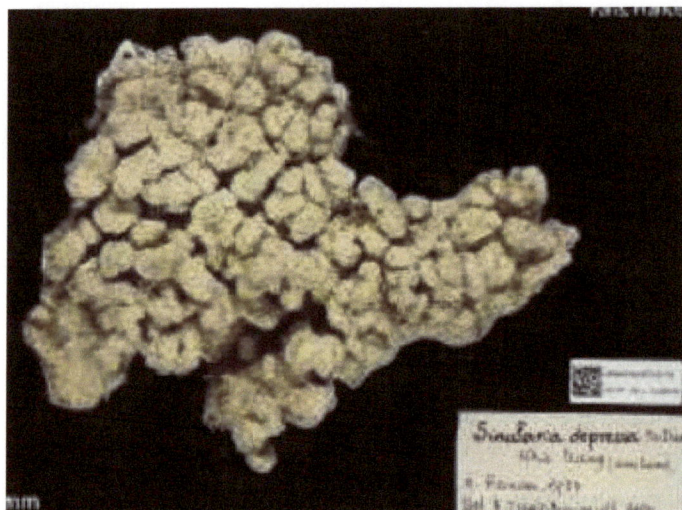

Fig. (129). Sterol 2, Sterol 3.

Sclerophytum erectum (= Sinularia erecta)

Global distribution: It is a common species of south equatorial East African reefs.

Ecology: This shallow coastal water species occurs commonly in a depth of about 10 m (Fig. **130**).

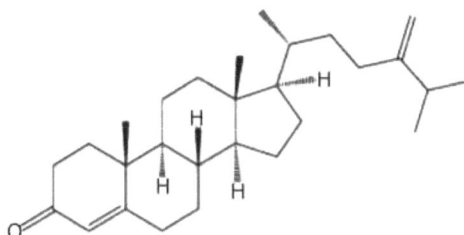

Fig. (130). *Sclerophytum erectum* (Marie HENNION; Creative Commons Attribution 4.0 International license. https://commons.wikimedia.org/wiki/File:Sinularia_(MNHN-IK-2015-2072)_001.jpeg.

Antidiabetics and their mechanisms of action: The methanol extracts of this cnidarian species exhibited antidiabetic effects by reducing plasma glucose levels in experimental rats [8].

Sclerophytum firmum (= Sinularia firma)

Global distribution: Indo-Pacific.

Ecology: It is a coral reef-based species.

Antidiabetics and their Mechanisms of action: The methanolic extracts of this species have shown antidiabetic activity by reducing plasma glucose levels in experimental rats [8].

Sclerophytum flexibile (= Sinularia flexibilis)

Global distribution: It occurs commonly in the Great Barrier Reef.

Ecology: It is often seen inhabiting shallow coastal areas where strong currents exist (Fig. **131**).

Antidiabetics and their mechanisms of action: The steroid 7α-hydroxy-y-crassarosterol A (Fig. **132**) derived from this species has shown weak PTP1B inhibitory activity (IC50 value of 33.05 µM) [2].

Scelrophytum molestum (Sinularia molesta)

Global distribution: Western Central Pacific: New Caledonia. Ecology: It is a shallow water and coral reef-associated species (Fig. **133**).

Antidiabetics and their mechanisms of action: A furanosesquiterpene and two guaiane-type sesquiterpenes (Figs. **134, 135**) isolated from this species have shown potent PTP1B inhibitory activities with IC50 values of 1.2, 218, and 344 µM, respectively [2].

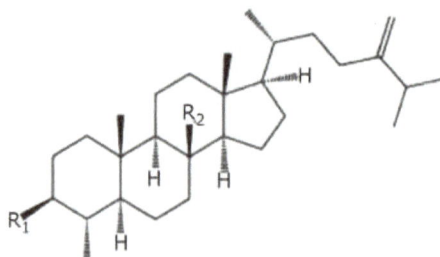

Sterol 2 (R1=OH: R2=OH)

Sterol 3 (R1= OCOOH3: R2=OH)

Fig. (131). *Sclerophytum flexibile* Albert kok; Creative Commons Attribution-Share Alike 4.0 International license. https://commons.wikimedia.org/wiki/File:Sinularia_flexibilis.jpg.

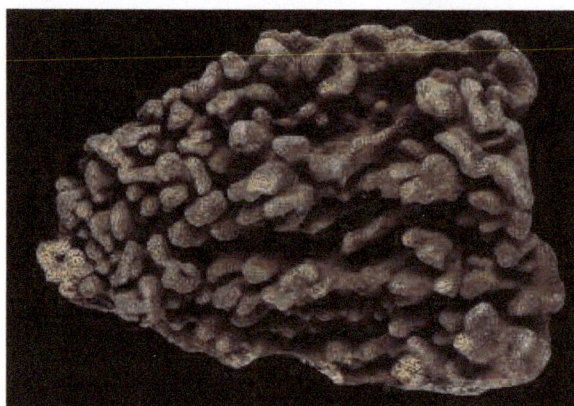

Fig. (132). 7α-hydroxy-crassarosterol A.

Fig. (133). *Scelrophytum molestum* Marie HENNION; Creative Commons Attribution 4.0 International license. https://commons.wikimedia.org/wiki/File:Sinularia_molesta_(MNHN-IK-2015-2067).jpeg.

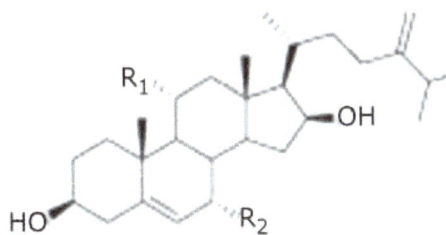

R1= CH3; R2= OH

Fig. (134). Furanosesquiterpene.

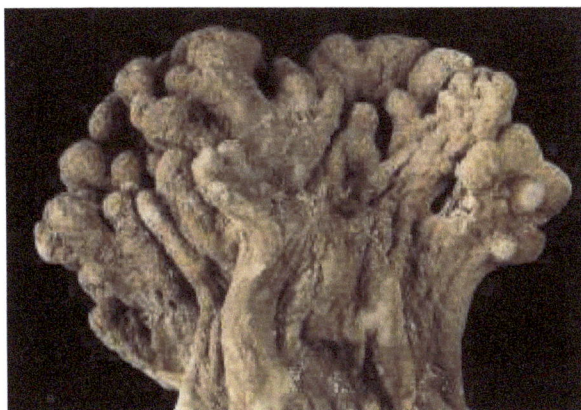

Fig. (135). guaiane-type sesquiterpenes.

Sclerophytum polydactylum (= Sinularia polydactyla)

Global distribution: It is common species of the Red Sea. Ecology: It is abundant in coral reef habitats (Fig. **136**).

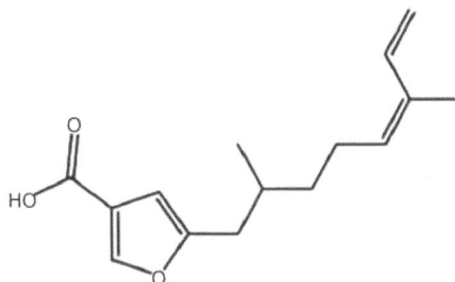

Fig. (136). *Sclerophytum polydactylum.* Derek Keats from Johannesburg, South Africa; Creative Commons Attribution 2.0 Generic license. https://commons.wikimedia.org/wiki/File:Finger_leather_coral,_Sinularia_ polydactyla_(6165871011).jpg.

Antidiabetics and their mechanisms of action: Two prenyleudesmane-type diterpenes (1,2) and one capnosane-type diterpenoid (Figs. **137-139**) isolated from this species have shown weak inhibitory activity against PTP1B with IC50 values of 75.5 μM, 63.9 μM, and 51.8 μM, respectively [2].

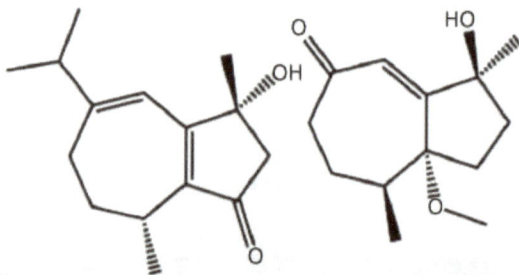

Fig. (137). Prenyleudesmane-type diterpene (1).

Fig. (138). Prenyleudesmane-type diterpene (2).

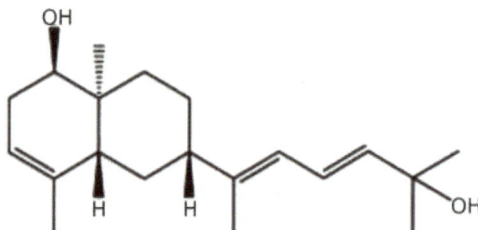

Fig. (139). capnosane-type diterpenoid.

Sinularia brassica

Global distribution: Western Pacific: New Caledonia, Palau, and Hong Kong. Ecology: It is commonly found at depths between 5 and 30m (Fig. **140**).

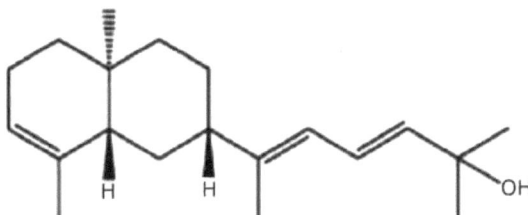

Fig. (140). *Sinularia brassica* . Neville Wootton; Creative Commons Attribution 2.0 Generic license. https://commons.wikimedia.org/wiki/File:Sinularia_brassica,_Maldivas.jpg 14.

Antidiabetics and their mechanisms of action: A furanocembranoid compound *viz.* epoxypukalide (Fig. **141**) isolated from this organism showed antidiabetic activity by enhancing the β-cell proliferation through the ERK1/2 signaling pathway activation and *via* the upregulations of cyclin D2 and E [6].

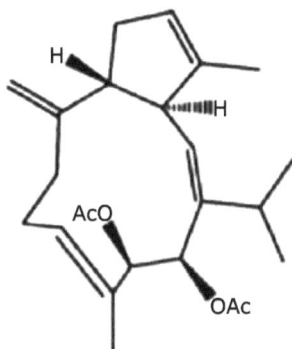

Fig. (141). Epoxypukalide.

Sinularia sp.

Antidiabetics and their mechanisms of action: A cembranoid *viz.* sinulin D and two terpenoids, namely 15-hydroxy-α-cadinol and (1R,3S,4S,7E,11E)-3, 4-epoxycembra-7,11,15-triene (Figs. **142-144**), have been derived from this species. Of these compounds, sinulin D and (1R,3S,4S,7E,11E)-3, 4-epoxycembra-7,11-15-triene have shown PTP1B inhibitory activity with IC50 values of 47.5, 22.1, and 12.5 mM, respectively [2].

Fig. (142). Sinulin D.

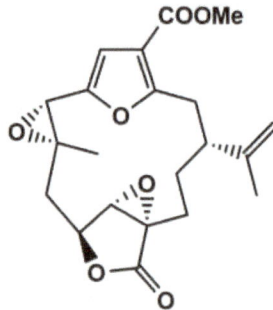

Fig. (143). 15-hydroxy-α-cadinol.

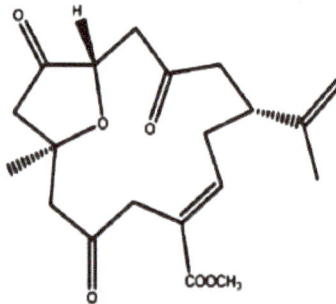

Fig. (144). (1R,3S,4S,7E,11E)-3,4-epoxycembra-7,11,15-triene.

Stichodactyla helianthus

Global distribution: Caribbean and Western Atlantic Seas.

Ecology: It is commonly seen in shallow areas with mild to strong currents (Fig. **145**).

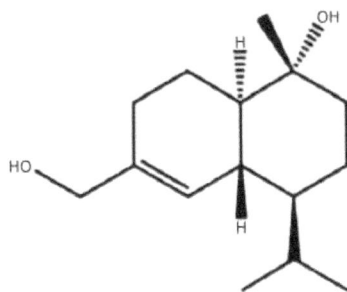

Fig. (145). *Stichodactyla helianthus.* Liné1 ; Creative Commons Attribution-Share Alike 3.0 Unported license. https://commons.wikimedia.org/wiki/File:Stoichactis_sp.jpg.

Antidiabetics and their mechanisms of action: A 44-residue peptide *viz.* helianthamide derived form this species has shown human salivary α-amylase inhibitory activity with a inhibition constant (Ki) value of 10 pM [2].

Annelid Worm

Urechis unicinctus

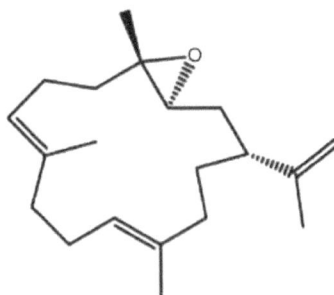

Fig. (146). Urechis unicinctus Christophe95 ; e Creative Commons Attribution-Share Alike 4.0 International license. https://commons.wikimedia.org/wiki/File:Urechis_unicinctus_in_Sokcho.jpg.

Global distribution: It is mainly distributed around the coasts of China, Korea, Russia, and Japan.

Ecology: It dwells in marine sediments of the lower intertidal and subtidal zones in coastal sandy and mud areas.

Antidiabetics and their mechanisms of action: A glycosaminoglycan isolated from this worm showed potent antidiabetic activity by reducing the blood glucose level and insulin resistance in diabetic rats [6] (Fig. **146**).

Crustaceans

Palaemon carinicauda (= Exopalaemon carinicauda)

Global distribution: It is known to occur in Bohai and Yellow Seas, China, and Singapore (Fig. **147**).

Ecology: It dwells in estuarine and coastal marine waters.

Fig. (147). *Palaemon carinicauda.* Laura Flamme; Creative Commons Attribution 4.0 International license. https://commons.wikimedia.org/wiki/File:Exopalaemon_annandalei_(MNHN-IU-2014-22826).jpeg.

Antidiabetics and their mechanisms of action: An insulin-like peptide that is similar to vertebrate insulin derived from this species has shown antidiabetic activity [98].

Penaeus japonicus (= Marsupenaeus japonicus)

Global distribution: Indo-West Pacific and Eastern Atlantic.

Ecology: It inhabits inshore areas up to 90 m deep on sandy mud bottoms.

Antidiabetics and their mechanisms of action: Four major types of insulin-like peptides, *viz.* insulin, relaxin, gonadulin, and androgenic gland hormone (AGH)/insulin-like androgenic gland factor (IAG), have been derived from this species. The antidiabetic effects of these peptides are, however, yet to be understood [99] (Fig. **148**).

Portunus trituberculatus

Global distribution: China, Korea, Japan, and Taiwan.

Ecology: It lives in sandy to sandy-muddy substrates in shallow coastal waters (Fig. **149**).

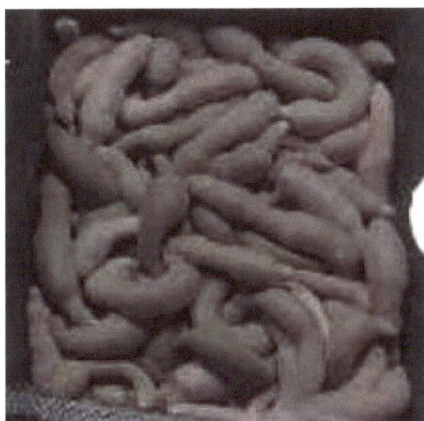

Fig. (148). *Penaeus japonicus* Daderot ; Creative Commons CC0 1.0 Universal Public Domain Dedication .; https://commons.wikimedia.org/wiki/File:Marsupenaeus_japonicus_National_Museum_of_Nature_and_Scie nce,_Tokyo_-_DSC07540.JPG.

Fig. (149). *Portunus trituberculatus, (.* NOAA; ublic domain; https://commons.wikimedia.org/wiki/File: Portunus_ trituberculatus.gif.

Antidiabetics and their mechanisms of action: An insulin-like androgenic gland hormone with antidiabetic activity has been derived from the androgenic gland of the male cab of this species. A short-term silencing of this hormone with RNA was found to decrease the transcript levels of insulin receptors and insulin-like growth factor-binding proteins, and this suggests that this hormone might perform its biological function through the insulin family-based signaling system [100].

Euphausia superba

Global distribution: It is seen only in the Southern Ocean.

Ecology: It mainly lives in oceans, seas, or other bodies of salt water.

Antidiabetics and their mechanisms of action: The peptides isolated from this krill species have shown antidiabetic activity by inhibiting ACE and DPP-IV [101] (Fig. **150**).

Fig. (150). *Euphausia superba.* Uwe Kils ; Creative Commons Attribution-Share Alike 3.0 Unported license. https://en.m.wikipedia.org/wiki/File:Antarctic_krill_%28Euphausia_superba%29.jpg.

Mollusks

Bivalves

Crassostrea madrasensis

Global distribution: Seas around Bangladesh, Pakistan, and India. Ecology: It is commonly seen in brackish waters and estuaries.

Antidiabetics and their mechanisms of action: The crude extracts of this species have been reported to yield α-glucosidase, α-amylase, and DPP-4 inhibitory activities [102] (Fig. **151**).

Fig. (151). *Crassostrea madrasensis* – Own.

Paphia malabarica

Global distribution: It is found distributed from the Gulf of Oman to Japan, including China, India, and the Philippines.

Ecology: It is found both in estuaries and coastal waters.

Antidiabetics and their mechanisms of action: Its ethyl acetate-methanol extracts containing phenol (88.62 mg GAE (gallic acid equivalence) /g) have shown α-glucosidase, α-amylase, and DPP-4 inhibitory activities with IC50 values of 1.5, 1.4, and 1.0 mg/mL, respectively [103] (Fig. **152**).

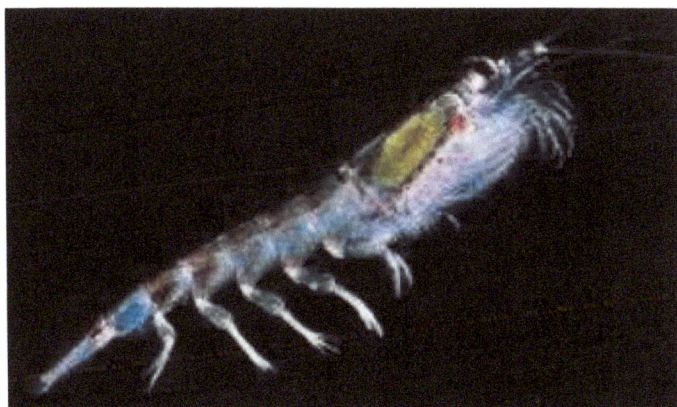

Fig. (152). *Paphia malabarica* Own.

Ruditapes philippinarum

Global distribution: It is native to the Pacific coast of Asia, and it is also seen in North America and Europe.

Ecology: It is most abundant in shallow areas with coarse sand and gravel.

Antidiabetics and their mechanisms of action: The peptides isolated from the extracts of this species have been reported to possess potent food components for preventing diabetes [101]. Its food protein-derived hydrolysates exhibited *in vitro* dipeptidyl peptidase (DPP-IV) inhibitory activity, and its papain had similar activity with an IC50 value of 0.7mg/ml [104] (Fig. **153**).

Fig. (153). *Ruditapes philippinarum.* H. Zell ; Creative Commons Attribution-Share Alike 3.0 Unported license. https://commons.wikimedia.org/wiki/File:Ruditapes_philippinarum_002.jpg.

Villorita cyprinoides

Global distribution: It is a common species of India.

Ecology: It is seen just below the surface of the soft bottom sediments.

Antidiabetics and their mechanisms of action: Its ethyl acetate-methanol extracts containing phenol (73.9 mg GAE (gallic acid equivalence) /g) have shown α-glucosidase, α-amylase, and DPP-4 inhibitory activities with IC50 values of 1.5, 1.5, and 1.1 mg/mL, respectively [103] (Fig. **154**).

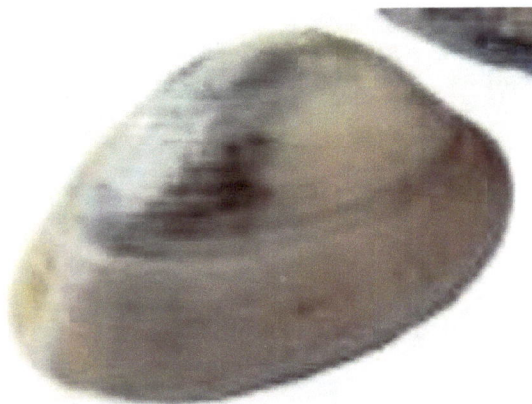

Fig. (154). *Villorita cyprinoides.* Public domain.; https://commons.wikimedia.org/wiki/File:Villorita _cyprinoides_002_in_an_instant_miso_soup.jpg.

Gastropods

Babylonia spirata

Global distribution: It is mainly distributed in the southeast coastal localities of India.

Ecology: It is known to inhabit both pelagic and benthic zones up to the depths of 60 m.

Antidiabetics and their mechanisms of action: The ethyl acetate-methanol extracts of this species displayed antidiabetic properties by inhibiting the carbolytic enzymes *viz.* α-amylase and α-glycosidase with IC50 values of 2.32 and 4.50 mg/mL, respectively [105] (Fig. **155**).

Fig. (155). *Babylonia spirata.* H. Zell ; Creative Commons Attribution-Share Alike 3.0 Unported license. https://commons.wikimedia.org/wiki/File:Babylonia_spirata_01.JPG.

Chicoreus ramosus

Global distribution: Indo-West Pacific: From east to South Africa, including Madagascar, Red Sea Mozambique, and Tanzania.

Ecology: It inhabits sandy and rubble bottoms near coral reefs to depths of about 10 m.

Antidiabetics and their mechanisms of action: The ethyl acetate-methanol extracts of this species displayed antidiabetic properties by inhibiting the carbolytic enzymes *viz.* α-amylase and α-glycosidase with IC50 values of 1.06 and 2.25 mg/mL, respectively [105] (Fig. **156**).

Fig. (156). *Chicoreus ramosus.* Didier Descouens ; Creative Commons Attribution-Share Alike 3.0 Unported license. https://en.m.wikipedia.org/wiki/File:Chicoreus_ramosus.png

Conus geographus

Global distribution: Tropical Indo-Pacific areas. Ecology: It inhabits coral reefs.

Antidiabetics and their mechanisms of action: The conopeptide *viz.* venom insulin, Con-Ins G1 (Cone Insulin G1) of this species has been reported to possess unique features for designing new insulin therapeutics. This venom insulin is active at the human insulin receptor [105], and it shows antidiabetic properties by inducing hypoglycemia with an IC50 value of 65ng/g [27] (Fig. **157**).

Fig. (157). *Conus geographus*. Almed2; he Creative Commons Attribution-Share Alike 3.0 Unported license. https://commons.wikimedia.org/wiki/File:Conus_geographus_4.jpg.

Turbinella pyrum

Global distribution: It is a common species of the Indian Ocean. Ecology: It inhabits the sandy bottom with mud and organic matter.

Antidiabetics and their mechanisms of action: The methanolic extracts of this species have shown antidiabetic activity by reducing the blood glucose level of diabetic rats [106]. The values of reduction during the different periods of the experiment are shown in Table **10** (Fig. **158**).

Table 10. Antidiabetic activity (as reduction of blood glucose level, mg/dl) of methanolic extracts of Turbinella pyrum in alloxan-induced diabetic rats during 0-15 days of experiment [106].

Extract	Dosage (mg/kg)	0 day	1st day	3rd day	5th day	6th day	7th day	15th day
Control	10	230.2	235.2	242.3	248.4	235.4	231.9	226.5
Methanolic extract	200	226.3	87.5	210.3	150.1	62.1	73.1	39.0
Methanolic extract	400	221.3	206.2	172.3	132.1	85.7	76.4	55.4

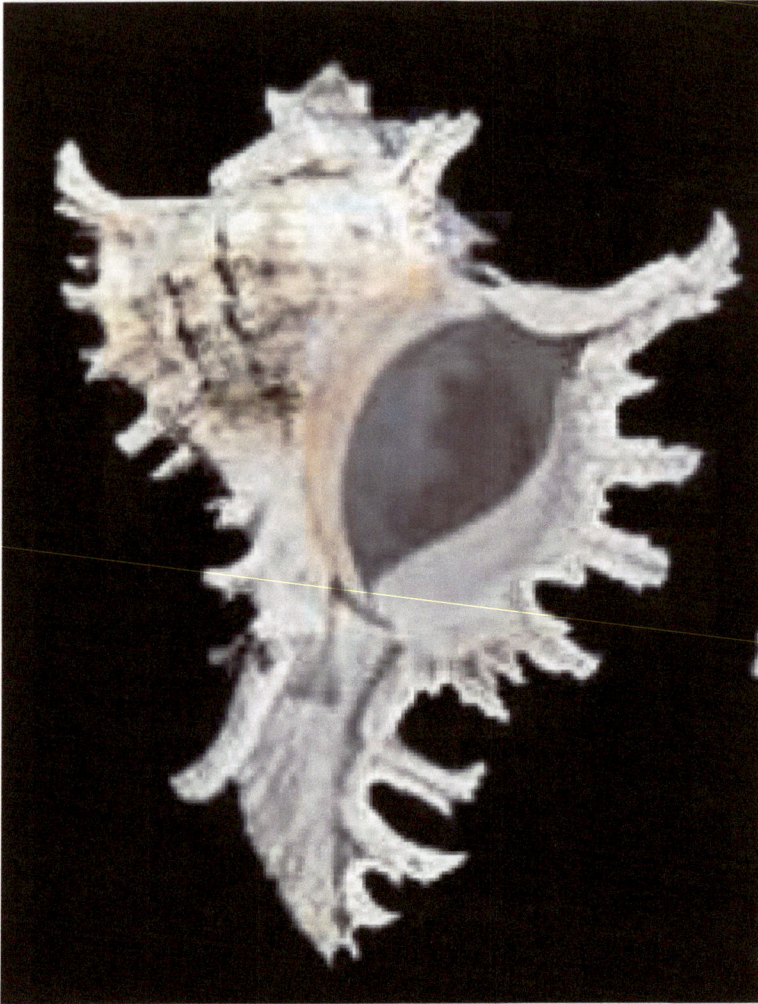

Fig. (158). *Turbinella pyrum.* Joop Trausel | Frans Slieker; Creative Commons Attribution-Share Alike 4.0 International license. https://commons.wikimedia.org/wiki/File:Turbinella_pyrum_NMR993000097157A. jpg.

Cephalopods

Acanthosepion pharaonis (= Sepia pharaonis)

Global distribution: It is widely distributed in the Indo-West Pacific, from the Red Sea to Japan and Australia (Fig. **159**).

Ecology: It lives in coral reefs with low nutrient availability.

Fig. (159). *Acanthosepion pharaonis* Portioid; Creative Commons Attribution-Share Alike 4.0 International license. https://commons.wikimedia.org/wiki/File:Acanthosepion_pharaonis.jpg.

Amphioctopus marginatus (= Octopus marginatus)

Global distribution: Tropical western Pacific Ocean.

Ecology: It is found in coastal muddy waters (Fig. **160**).

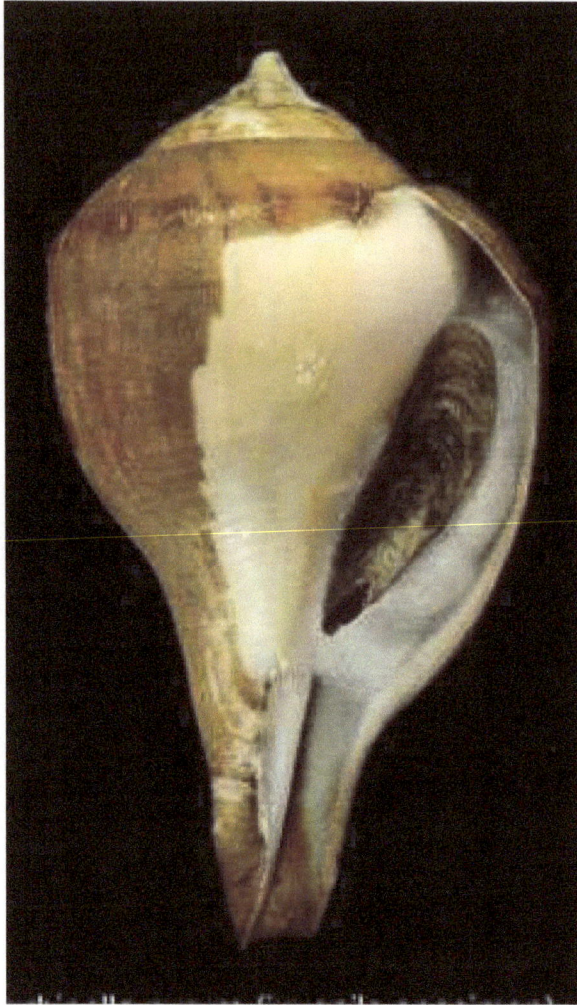

Fig. (160). *Amphioctopus marginatus*. Nick Hobgood; Creative Commons Attribution-Share Alike 3.0 Unported license. https://commons.wikimedia.org/wiki/File:Octopus_shell.jpg.

Cistopus indicus (= Octopus indicus)

Global distribution: This globally distributed species is commonly seen in the temperate and subtropical seas.

Ecology: It is a benthic species that occurs on the mud bottom with depths ranging from 0 to 50 m (Fig. **161**).

Fig. (161). *Cistopus indicus.* Image credit: JIAN-XIANG LIAO1 AND CHUNG-CHENG LU . (Reproduced with kind permission). source: jian-xiang liao and chung-cheng lu. a new species of cistopus (cephalopoda: octopodidae) from taiwan and morphology of mucous pouches. journal of molluscan studies (2009) 75: 269–278 (fig. **8a**) (to be).

Sepiella inermis

Global distribution: Throughout the Indo-Pacific.

Ecology: It is known to occur from shallow water down to depths of about 40m (Fig. **162**).

Fig. (162). *Sepiella inermis* Image credit: Soumya Krishnan, Kajal Chakraborty, and Minju Joy. Source: Soumya Krishnan, Kajal Chakraborty, and Minju Joy. (2019). First report of chromenyl derivatives from spineless marine cuttlefish Sepiella inermis: Prospective antihyperglycemic agents attenuate serine protease dipeptidyl peptidase IV. J.Food Biochem. DOI: 10.1111/jfbc .12824 (Fig. **a**)

Antidiabetics and their mechanisms of action: The ethyl acetate-methanol extracts of, *Amphioctopus marginatus, Cistopus indicus, Sepiella inermis,* and *Uroteuthis (Photololigo) duvaucelii* have shown alpha-glucosidase, alpha-amylase, and DPP-4 inhibitory activities [107], and IC50 values recorded for the different species are given in Table **11**.

Table 11. Alpha-glucosidase, alpha-amylase, and DPP-4 inhibitory activities ethyl acetate-methanol extracts [107].

Species	For Alpha-glucosidase (IC50,mg/mL)	For Alpha-amylase (IC50,mg/mL)	For DPP-4 (IC50,mg/mL)
Acanthosepion pharaonic	5.37	1.96	5.37
Amphioctopus marginatus	2.79	2.50	3.60
Cistopus indicus	2.83	1.69	2.51
Sepiella inermis	2.42	1.76	3.35
Uroteuthis (Photololigo) duvaucelii	4.75	1.71	4.54

Uroteuthis (Photololigo) duvaucelii (= Urothethis duvaucelii)

Global distribution: Indo-Pacific ocean: Red Sea and Arabian Sea; from Mozambique to South China and Philippines Sea; and Taiwan (Fig. **163**).

Ecology: It is seen at depths between 3–170 m.

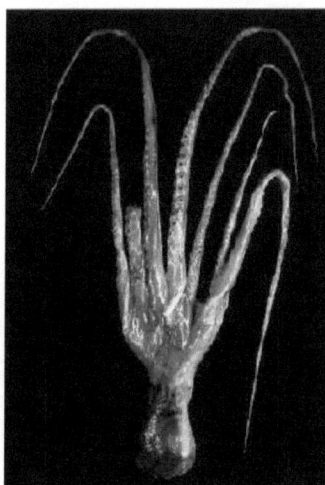

Fig. (163). *Urothethis duvauceli* A. Pollock; Public domain. https://commons .wikimedia.org/wiki/File: Uroteuthis_duvauceli.jpg.

Antidiabetics and their mechanisms of action: Two chromenyl derivatives *viz.* 11-(3,4,4α,5,8,8α-hexahydro-8-methoxy-4-methyl-1H-isochromen-4-yloxy)-11-hydroxyethylpentanoate (Fig. **164**) and methyl 9-(4,4α,5,8-tetrahydro -3-oxo-3H-isochromen-5-yl)hexanoate (Fig. **165**) isolated from this species showed antidiabetic activity by inhibiting α-amylase, α-glucosidase, and DPP-IV enzymes. Compound 1 registered IC50 values for these enzymes with IC50 values of 0.27, 0.19, and 0.16 mg/ml, respectively, and the corresponding values for compound 2 were 0.36, 0.28, and 0.23 mg/ml, respectively [6, 108].

Fig. (164). 11-(3,4,4α,5,8,8α-hexahydro-8-methoxy-4-methyl-1H-isochromenyloxy)-11-hydroxyethyl-pentanoate.

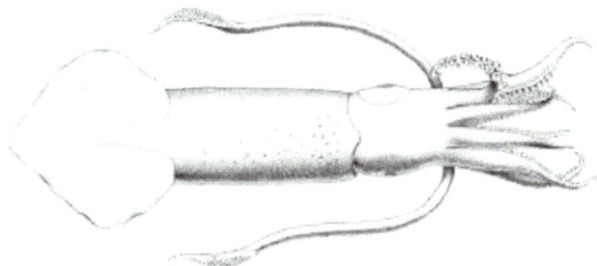

Fig. (165). Methyl 9-(4,4α,5,8-tetrahydro-3-oxo-3H-isochromen-5-yl)hexanoate.

Others

Squid pen (Gladius)

Antidiabetics and their mechanisms of action: The phenolic acid, homogentisic acid (Fig. **167**), derived from the squid pen (or gladius, the hard internal body part) (Fig. **166**) by bacterial fermentation with the species of Paenibacillus (non-sugar-based α-glucosidase inhibitor) exhibited higher α-glucosidase activity [6].

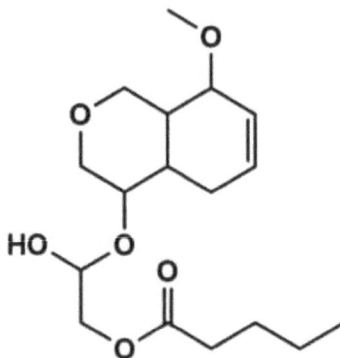

Fig. (166). Squid pen Pen (gladius) of a squid (*Sepioteuthis* sp.).

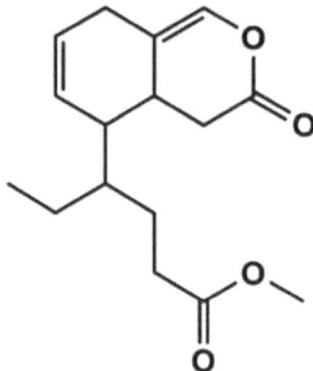

Fig. (167). Homogentisic acid.

Echinoderms

Starfish

Acanthaster planci

Global distribution: Indian Ocean and Pacific Ocean. Ecology: It is commonly found in coral reefs (Fig. **168**).

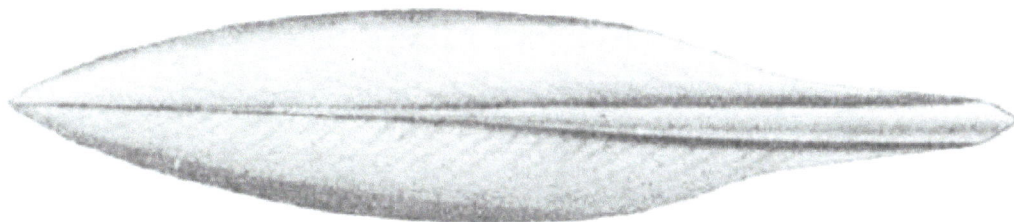

Fig. (168). *Acanthaster planci* . Kris Mikael Krister Creative Commons Attribution 3.0 Unported license. https://commons.wikimedia.org/wiki/File:Crown_Of_Thorns_Starfish_Acanthaster_Planci_(223180421).jpeg

Antidiabetics and their mechanisms of action: Two steroids *viz.* 5α-cholesta-24 en-3β,20β-diol-23-one and 5α-cholesta-9(11)-en-3β, 20β-diol (Figs. **169, 170**) possessed antidiabetic effects by inhibiting anti-α-glucosidase with IC50 values of 58 and 55 0.5 μg/ml [109].

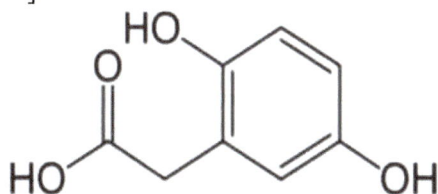

Fig. (169). 5α-cholesta-24-en-3β,20β-diol-23-one.

Fig. (170). 5α-cholesta-9(11)-en-3β, 20.

Asterias rollestoni

Global distribution: In the seas around Japan and the Yellow Sea.

Ecology: It is found in the littoral zone (Fig. **171**).

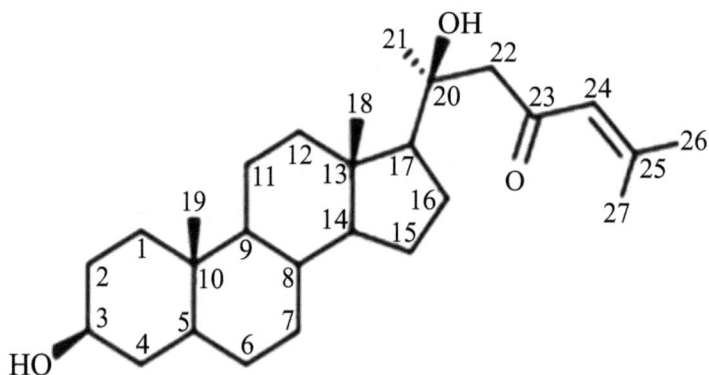

Fig. (171). *Asterias rollestoni* (Daniel J. Drew; Creative Commons CC0 1.0 Universal Public Domain Dedication . https://commons.wikimedia.org/wiki/File:Asterias_rollestoni_(YPM_IZ_080407)_02.jpg

Antidiabetics and their mechanisms of action: At a concentration of 0.12mg/ml, the enolic saccharide *viz.* asterolloside (Fig. **172**) derived from this organism showed moderate alpha-glucosidase inhibitory activity with a percentage value of 37.9% [110].

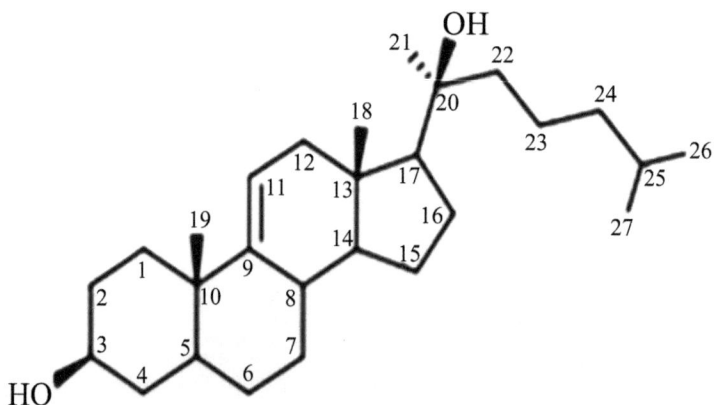

Fig. (172). Asterolloside.

Astropecten irregularis

Global distribution: It is commonly found on the west coast of Norway and Morocco.

Ecology: It is usually seen on sand or sandy mud, where it may be buried just below the surface (Fig. **173**).

Antidiabetics and their mechanisms of action: vide *Ophiura albida.*

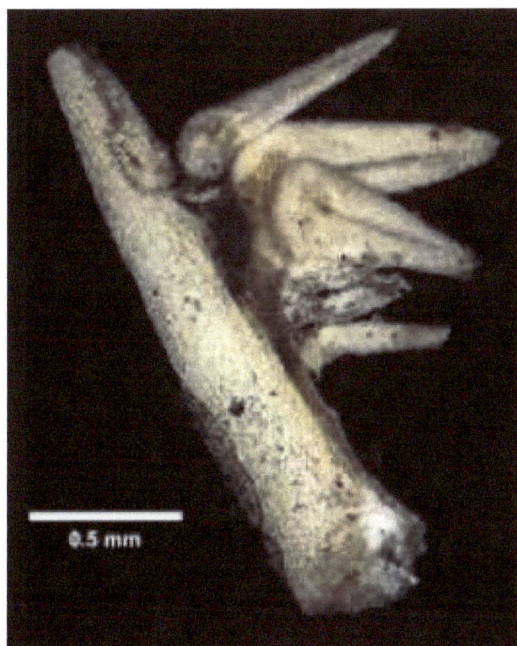

Fig. (173). *Astropecten irregularis,* . Espen Rekdal; Creative Commons Attribution 4.0 International license. https://commons.wikimedia.org/wiki/File:Astropecten_irregularis_01_Espen_Rekdal.jpg.

Luidia sarsii

Global distribution: It occurs in areas from Norway to the Mediterranean Sea. Ecology: It is found in deeper waters at a depth of 20 m (Fig. **174**).

Antidiabetics and their mechanisms of action: vide *Ophiura albida.*

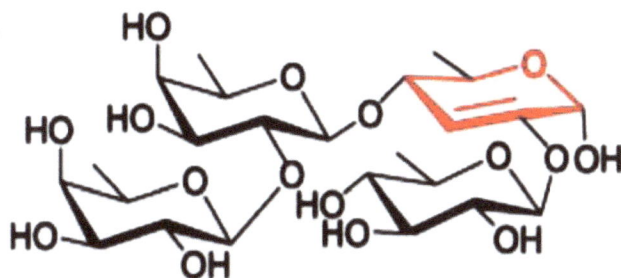

Fig. (174). *Luidia sarsi* Espen Rekdal ; Creative Commons Attribution 4.0 International license. https://commons.wikimedia.org/wiki/File:Luidia_sarsi_01_Espen_Rekdal.jpg.

Brittle Star

Ophiura albida

Global distribution: Mediterranean Sea and NE Atlantic.

Ecology: This benthic species is seen in soft substrates such as muddy sand, coarse sand, and fine sand at depths of 200-850m (Fig. **175**).

Fig. (175). *Ophiura albida.* Hans Hillewaert ; Creative Commons Attribution-Share Alike 4.0 International license. https://commons.wikimedia.org/wiki/File:Ophiura_albida.jpg.

Antidiabetics and their mechanisms of action: The extracts of*Astropecten irregularis, Luidia sarsii, and Ophiura albida* (brittle star) containing phenolic compounds such as pyrogallol, gallic, sinapic, ferulic, p-hydroxybenzoic and salicylic acids have been reported to exhibit alpha-amylase and alpha-glucosidase inhibitory activities [111], and the IC40 values registered for the different species are shown in Table **12**.

Table 12. Alpha-amylase and alpha-glucosidase inhibitory activities of the extracts of *Astropecten irregularis, Luidia sarsii,* and *Ophiura albida* [111].

Species	Alpha-amylase inhibition (IC50, µg/ml)	Alpha-glucosidase inhibition (IC50, µg/ml)
Astropecten irregularis	147.1	540.0
Luidia sarsii	150.5	442.8
Ophiura albida	737.3	872.3
Acarbose (control)	396.4	199.5

Sea Urchins

Echinometra mathaei

Global distribution: Tropical Indo-Pacific.

Ecology: It is found mainly on coral reefs at depths of 140 m (Fig. **176**).

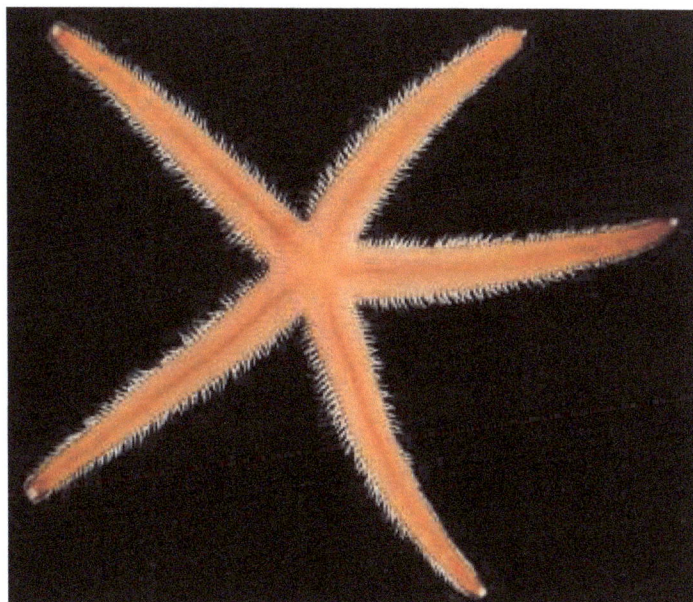

Fig. (176). *Echinometra mathaei* Gimp Savvy or from NOAA:; public domain ; https://commons.wikimedia.org/wiki/File:Seaurchin.jpg.

Antidiabetics and their mechanisms of action: The ethyl acetate extracts of the shell, gonad, and Aristotle's lantern of this species have shown α-glucosidase inhibitory activity [112]. The IC50 values recorded for the different organs of this species are shown in Table **13**.

Table 13. α-glucosidase inhibitory activity of ethyl acetate extracts of *Echinometra mathaei* [112].

Conc. of extract (%)	Shell (IC50, mg/mL)	Gonad (IC50, mg/mL)	Aristotle's lantern (IC50, mg/mL)
2.5	46.3	36.1	36.0
5	51.4	48.1	52.1
10	59.8	50.0	64.0
20	67.1	66.2	79.1
40	78.2	81.3	86.0

Glyptocidaris crenularis

Global distribution: Northern China.

Ecology: It inhabits mainly subtidal coastal areas (Fig. **177**).

Fig. (177). *Glyptocidaris crenularis*. GFDL; Creative Commons Attribution-Share Alike 3.0 Unported license. https://commons.wikimedia.org/wiki/File:Glyptocidaris_crenularis.jpg.

Antidiabetics and their mechanisms of action: The oxidized sterol, 5α,8α-Epidioxycholest-6-en-3β-ol (Fig. **178**), derived from this species, is under investigation for its antidiabetic activity potential [24].

Fig. (178). 5α,8α-Epidioxycholest-6-en-3β-ol.

Paracentrotus lividus

Global distribution: Throughout NE Atlantic and Mediterranean Sea. Ecology: It dwells in shallow subtidal rocky habitats.

Antidiabetics and their mechanisms of action: The gonad and intestine extracts of this species have shown antidiabetic activity by reducing the blood glucose concentrations in rats with DMT1 and DMT2, and the recorded values were 138.67 and 180.33 mg/dL, respectively [113] (Fig. **179**).

Fig. (179). *Paracentrotus lividus*. Frédéric Ducarme Creative Commons Attribution-Share Alike 4.0 International license. https://commons.wikimedia.org/wiki/File:Paracentrotus_lividus_Banyuls_2.jpg.

Scaphechinus mirabilis

Global distribution: North Pacific.

Ecology: In shallow marine sediments at depths of 0.5-125m.

Antidiabetics and their mechanisms of action: The water extracts from viscera and spines of this species have shown *in vitro* α-amylase enzyme reducing activity with values ranging from 19.41-42.46% [113] (Fig. **180**).

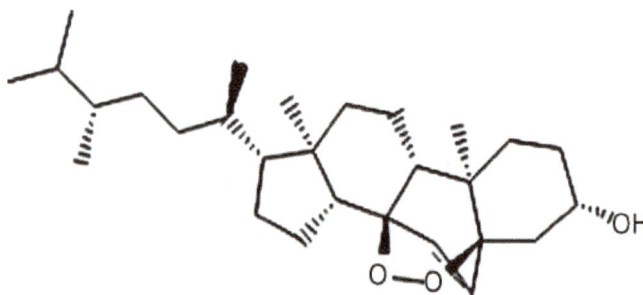

Fig. (180). *Scaphechinus mirabilis* . Harum.koh ; e Creative Commons Attribution-Share Alike 4.0 International license. https://commons.wikimedia.org/wiki/File:Scaphechinus_mirabilis_2090690.jpg.

Stomopneustes variolaris

Global distribution: Tropical and subtropical Indo-Pacific.

Ecology: It is commonly seen in rock pools, overhangs, crevices, and bores. It is known to prefer shady areas with constant water circulation.

Antidiabetics and their mechanisms of action: The extracts derived from the spine and tissue of this species have shown *in-vitro* α-amylase inhibitory activity with percentage values of 51,11 and 41.81, respectively [114] (Fig. **181**).

Fig. (181). *Stomopneustes variolaris* . FredD ; Creative Commons Attribution-Share Alike 4.0 International license. https://commons.wikimedia.org/wiki/File:Stomopneustes_variolaris.JPG.

Strongylocentrotus droebachiensis

Global distribution: It is commonly seen in northern waters of the world, including both the Pacific and Atlantic Oceans.

Ecology: It is found on rocky grounds down to depths of 1200 m.

Antidiabetics and their mechanisms of action: At a concentration of 25 µg/mL, the hexane extract of this species showed α-amylase inhibitory activity with a percentage value of 95.75 [115]. (Gelani *et al.*, 2022). A bioactive dark-red pigment *viz.* echinochrome A (7-ethyl-2,3,5,6,8-pentahydroxy- 1,4-naphthoquinone) derived from the shells and spines of this species has shown antidiabetic activity with improved glucose tolerance and potential reno-protective mechanism in a mouse model of type 2 diabetes. Further, this echinochrome A has been registered as a medicinal product and approved for medicinal use in Russia (reg. no. P N002363/01) and is the active substance of the drug Histochrome®, which has been approved for medicinal use in Russia against various diseases. Furthermore, this compound has the potential to provide a new therapeutic strategy for DN.- diabetic nephropathy [38] (Fig. **182**).

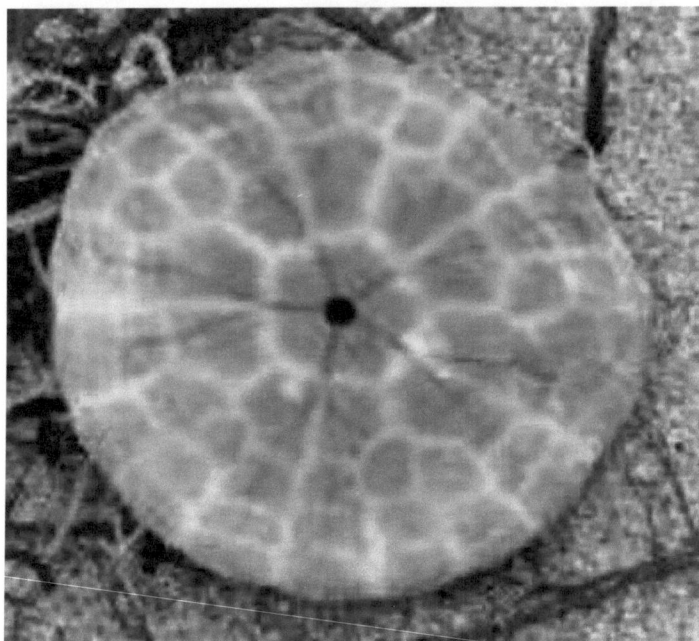

Fig. (182). *Strongylocentrotus droebachiensis* . Hannah Robinson Creative Commons CC0 License ; https://commons.wikimedia.org/wiki/File:Underside.JPG.

Sea Lilies

Comaster schlegelii

Global distribution: Indonesia, Fiji, and Japan, as well as Maldives and Papua New Guinea.

Ecology: It is most common in shallow waters on reef crests down to a depth of about 5 m.

Antidiabetics and their mechanisms of action: The extracts of this species containing fatty acids and sterols have been reported to possess a moderate inhibitory effect on α-amylase enzyme, which is commonly exploited as a drug target for preventing postprandial hyperglycemia in diabetes [116] (Fig. **183**).

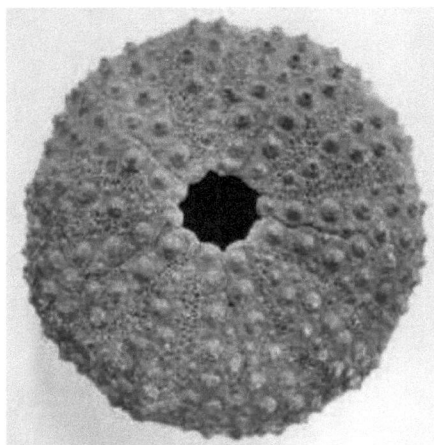

Fig. (183). *Comaster schlegelii.* Anne Hoggett; Creative Commons Attribution 3.0 Unported license. https://commons.wikimedia.org/wiki/File:Comaster_schlegelii_-_South_Island.jpeg.

Himerometra robustipinna

Global distribution: It is widespread in the western Pacific and Indian Ocean.

Ecology: It inhabits coastal coral reef ecosystems influenced by strong currents at depths of 0 – 60 m.

Antidiabetics and their mechanisms of action: The extracts of this species containing fatty acids and sterols have been reported to possess a moderate inhibitory effect on α-amylase enzyme, which is commonly exploited as a drug target for preventing postprandial hyperglycemia in diabetes [116] (Fig. **184**).

Sea Cucumbers

Global distribution: Western Pacific: Palau, Micronesia, Tonga, Australia, and New Caledonia.

Ecology: It lives in reef ecosystems at depths up to 30 m, and it is occasionally found in inshore reefs and lagoons.

Antidiabetics and their mechanisms of action: The methanol extracts of this species have shown maximum α-amylase inhibition with a percentage value of 91.70 [115] (Fig. **185**).

Fig. (184). *Himerometra robustipinna*. Holobionics Creative Commons Attribution-Share Alike 4.0 International license. https://commons.wikimedia.org/wiki/File:Barren_Island_feather_star_and_branching_ coral.jpg.

Fig. (185). *Actinopyga* sp. Julien Bidet Creative Commons Attribution-Share Alike 4.0 International license. https://commons.wikimedia.org/wiki/File:Actinopyga_miliaris.JPG.

Apostichopus japonicus(=Stichopus japonicus)

Global distribution: Southeast Asia.

Ecology: It is found in shallow waters along the coasts (Fig. **186**).

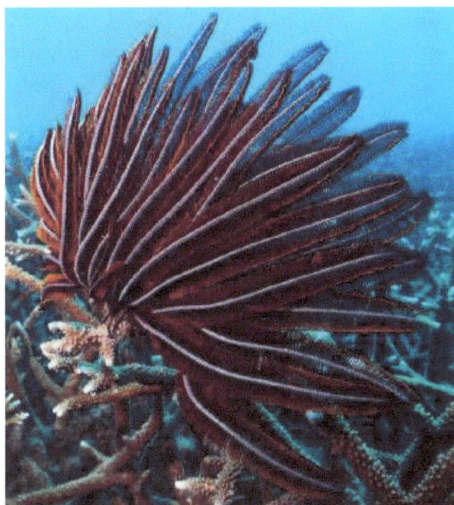

Fig. (186). *Apostichopus* sp.; Dr. Bradley Stevens and Eric Munk. NOAA; Creative Commons CC0 License; https://commons.wikimedia.org/wiki/File:Parastichopus_californicus.jpg.

Antidiabetics and their mechanisms of action: The vanadium-containing proteins (VCPs) derived from this species have been reported to decrease fasting blood glucose and serum insulin levels and manage insulin resistance. High-fat, high-sugar diet mice treated with VCP showed reduced fasting blood glucose and increased insulin sensitivity [6]. These authors further stated that compounds 1,3-dipalmitolein (Fig. **187**) and cis-9-octadecenoic acid (Fig. **188**) showed alpha-glucosidase inhibitory activity. Its two unsaturated fatty acids *viz.* 7(Z)-octadecenoic acid (Fig. **189**) and 7(Z),10(Z)-octadecadienoic acid (Fig. **190**) inhibited Bacillus stearothermophilus α- glucosidase with IC50 values of 0.49 and 0.60 μg mL−1, respectively, and inhibited Saccharomyces cerevisiae α-glucosidase with IC50 values of 0.51 and 0.67 μg mL−1, respectively [2] . The glycosaminoglycan derived from this species showed a hypoglycemic effect besides suppressing hepatic glucose production in insulin-resistant hepatocytes in experimental animals treated at 50 - 100 mg/kg/day [117]. The compounds isopimarane diterpene (1), deoxydiaporthein A (2), and iso-pimara-8(14),15-diene (3) (Fig. **191**) derived from the fungus Epicoccum sp. associated with this sea cucumber displayed α-glucosidase inhibitory activity with IC50 values of 4.6 and 11.9 μM, respectively [118].

Fig. (187). 1,3-dipalmitolein.

Fig. (188). cis-9-octadecenoic acid.

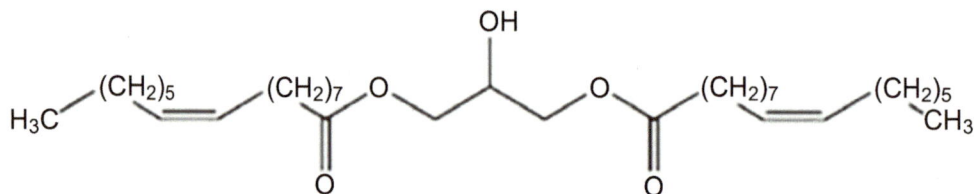

Fig. (189). 7(Z)-octadecenoic acid.

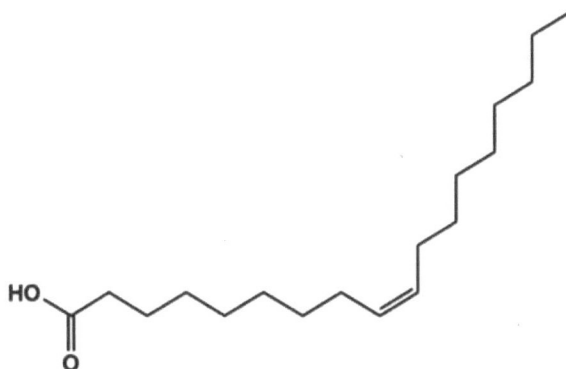

Fig. (190). 7(Z),10(Z)-octadecadienoic acid.

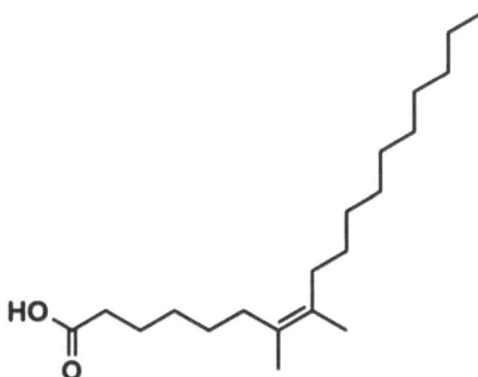

Fig. (191). Isopimarane diterpene — Deoxydiaporthein A — Iso-pimara-8(14),15-diene.

Bohadschia argus

Global distribution: Tropical eastern Indian Ocean and western Pacific.

Ecology: It is commonly found in and around coral reefs with exposed sandy seabed at a depth range of 2 -10 m.

Antidiabetics and their mechanisms of action: At 25-100 ppm concentrations, the extracts of this showed alpha-amylase inhibitory activity by delaying the absorption of starch into the body *via* blocking the hydrolysis of 1,4-glycosidic linkages of starch and other oligosaccharides into maltose, maltotriose, and other simple sugars [119] (Fig. **192**).

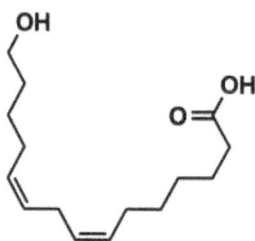

Fig. (192). *Bohadschia argus* Dan Schofield Creative Commons Attribution 4.0 International license.
https://commons.wikimedia.org/wiki/File:Bohadschia_argus_42214534.jpg

Cucumaria frondosa

Global distribution: North, Eastern, and Western Atlantic.

Ecology: It is seen in rocky areas of shallow water, and it is often found just below the low tide mark.

Antidiabetics and their mechanisms of action: The gonad hydrolysates of this species possessed antidiabetic activity by significantly reducing oral glucose tolerance, fasting blood glucose level, and insulin resistance in a rat model of type II diabetes induced by combined with a high-fat diet [120] (Fig. **193**).

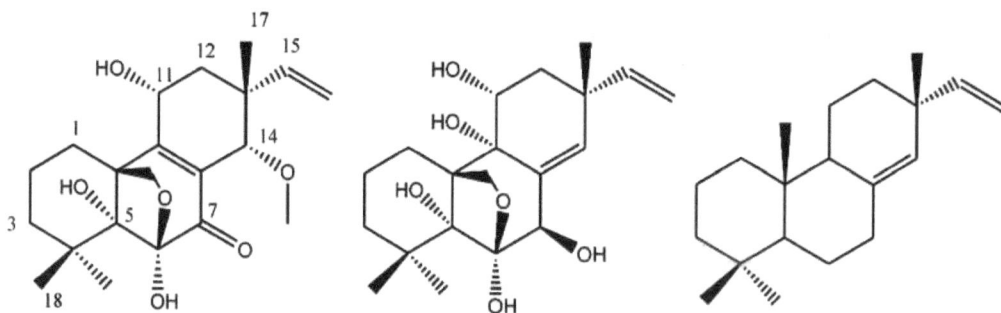

Isopimarane diterpene   Deoxydiaporthein A   Iso-pimara-8(14),15-diene

Fig. (193). *Cucumaria frondosa* Vsevolod Creative Commons Attribution 4.0 International license.
https://commons.wikimedia.org/wiki/File:Cucumaria_frondosa_184708776_2.jpg.

Holothuria (Thymiosycia) thomasi (= Holothuria thomasi)

Global distribution: Caribbean Sea and Gulf of Mexico.

Ecology: It inhabits mainly coral reefs at a depth range of 3-30 m (Fig. **194**).

Fig. (194). *Holothuria thomasi* HEROES DIVE ; Creative Commons Attribution-Share Alike 4.0 International license. https://commons.wikimedia.org/wiki/File:Holothuria_thomasi_Roatan.jpg.

Antidiabetics and their mechanisms of action: The saponin present in this species has been reported to show intestinal α-glucosidase and pancreatic α-amylase inhibitory activity [6] (Fig. **195**).

Fig. (195). Saponin.

CONCLUSION

Among the marine invertebrates, bioprospecting investigations have been done to a greater extent in marine sponges only. Such studies on free-living annelid worms, bryozoans, and crustaceans are either nil or very much limited. As these invertebrates with several phyla may possess novel bioactive compounds, intensive investigations on the chemical diversity and its therapeutic applications are the need of the hour.

CHAPTER 6

Antidiabetic Properties of Marine Chordates

Abstract: The antidiabetic properties of the natural products derived from tunicates, marine elasmobranchs, teleost fishes, and soft-shelled turtles are dealt with in this chapter. The global distribution and ecology of the antidiabetic species and the modes of action of their bioactive compounds are also depicted.

Keywords: Antidiabetic properties, Antidiabetic molecules, Distribution, Ecology, Marine fishes, Soft-shelled turtle, Tunicates.

INTRODUCTION

The marine chordates are the least studied group as far as chemical diversity in general and antidiabetic properties in particular are considered. Among the different components of marine chordates, chemical diversity has been fairly well studied in tunicates. While marine fish are the least studied group. No bioprospecting investigations are available on marine reptiles and mammals. Among the tunicates, several species are known to possess natural products with a wide range of bioactive properties like anticancer and antimicrobial. Very few species of this group showed antidiabetic properties. For example, the lipids extracted from the sea pineapple Halocynthia roretzi demonstrated antidiabetic properties in mice models. Marine fish are known to be important nutrient-rich foods possessing positive properties in the reduction of type 2 diabetes mellitus. The antidiabetic compounds of marine fish, such as proteins, peptides, and ω-3 Polyunsaturated Fatty Acids (PUFAs), have been reported to be potential sources in the prevention and management of T2DM.

Tunicates

Aplidium elegans (= Sidnyum elegans)

Global distribution: Atlantic Ocean, Mediterranean Sea, and English Channel.

Ecology: It dwells on coral reefs, hiding at the bases of corals at depths of 3-30 m.

Antidiabetics and their mechanisms of action: A monophosphorylated polyketide *viz.* phosphoeleganin (Fig. **1**) derived from this species showed PTP1B inhibitory effects with an IC50 value of 1.3M. Further, this compound was also found to inhibit the aldose reductase enzyme [2].

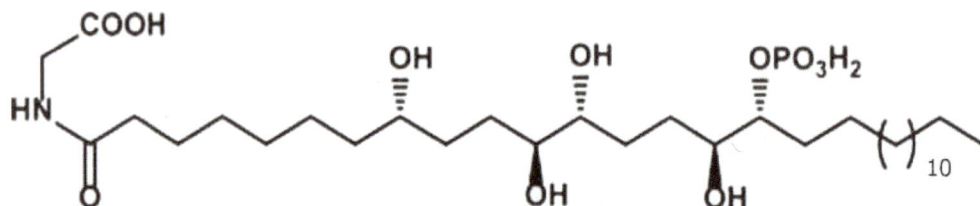

Fig. (1). Phosphoeleganin.

Ascidia ahodori

Global distribution: It is found widely distributed in polar, tropical, and temperate environments.

Ecology: This benthic species is seen on shores and littoral or intertidal areas.

Antidiabetics and their mechanisms of action: The methanol and ethyl acetate extracts of this species have shown antidiabetic activity by inhibiting the enzyme α-amylase. At the concentrations of 100, 200, and 300 µg/ml, the methanol extracts showed the percentage values of inhibition as 44, 63, and 71, respectively, and the corresponding values for the ethyl acetate extracts were nil, 29, and 45, respectively [121] (Fig. **2**).

Fig. (2). *Ascidia* sp. Gronk; Creative Commons Attribution-Share Alike 3.0 Unported, 2.5 Generic, 2.0 Generic and 1.0 Generic license.; https://commons.wikimedia.org/wiki/File:Ascidia.JPG.

Ascidia sp.

Antidiabetics and their mechanisms of action: The ethyl acetate extracts of this species have shown antidiabetic activity by inhibiting the enzyme α-amylase. At the concentrations of 100, 200, and 300 µg/ml, these extracts showed percentage values of inhibition as 44, 52, and 58, respectively [121].

Didemnum vexillum

Global distribution: North America, Europe, and New Zealand.

Ecology: It grows on seagrass and rock surfaces, and it also grows as a fouling organism on cultivated bivalves, net cages, and other man-made structures.

Antidiabetics and their mechanisms of action: The methanol and ethyl acetate extracts of this species have shown antidiabetic activity by inhibiting the enzyme α-amylase. At the methanol concentrations of 100, 200, and 300 µg/ml, the inhibition percentage values were 48, 62, and 74, respectively, and the corresponding values for the ethyl acetate extracts were 48, 54, and 61, respectively [121] (Fig. **3**).

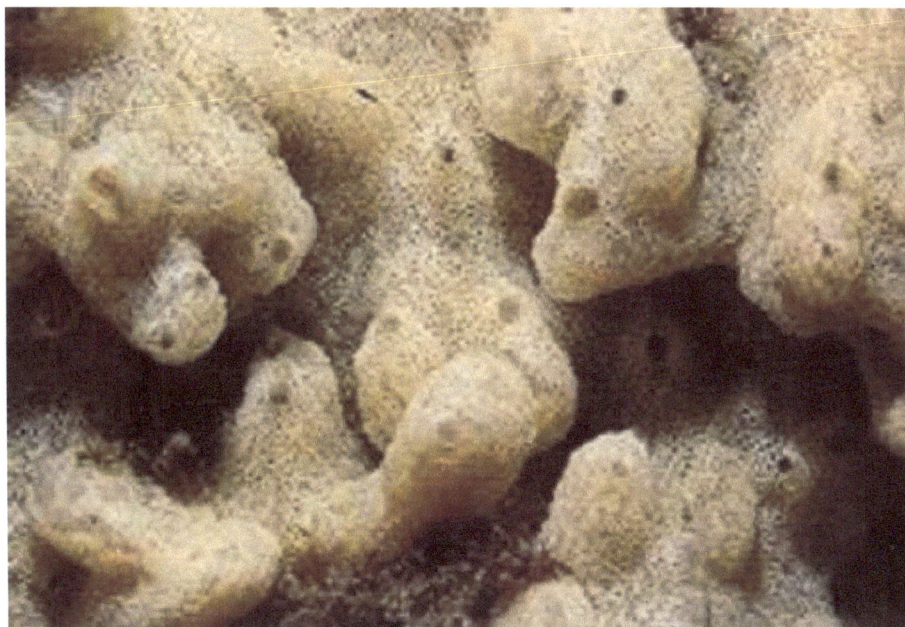

Fig. (3). *Didemnum vexillum* U.S. Geological Survey/photo by Dann Blackwood (USGS); public domain; https://commons.wikimedia.org/wiki/File:Tunicate_colony_of_Didemnum_vexillum.JPG.

Halocynthia roretzi

Global distribution: NW Pacific: Japan and Korea.

Ecology: It is commonly found in shallow coastal waters (Fig. **4**).

Fig. (4). *Halocynthia roretzi.*Yoshiaki Sano from Ichikawa, Chiba, Japan; Creative Commons Attribution-Share Alike 2.0 Generic license. https://commons.wikimedia.org/wiki/File:29017702_95d79d4124_o_d.jpg.

Antidiabetics and their mechanisms of action: A total of 11 active compounds, *viz.* gemfibrozil, daidzein, hesperetin, clofibrate, pravastatin, nateglinide, allatoin, trigonelline, L-arabinose, metformin, and astaxanthin (Figs. **5-8**) have been derived from the metabolome of this species. Of these compounds, daidzein, hesperetin, L-arabinose, trigonelline, and astaxanthin were found to improve dyslipidemia, hyperglycemia, and other complications associated with diabetes mellitus. Further, oral administration of its dry powder at a dose of 125 mg/kg resulted in a significant reduction in blood glucose levels, and the blood glucose levels were 12.21 mmol/L and 8.1 mmol/L after 0.5 h and 2 h of glucose gavage, respectively. Furthermore, metformin and nateglinide, the two first-line drugs for the treatment of type 2 diabetes, have been detected in the tissue of this species.

Metformin has been reported to possess beneficial effects on glycemic control during systemic glucocorticoid treatment [1222]. The lipid extracts of this species have shown hypoglycemic effects in diabetic mice, and its two amino acid derivatives *viz*. Herdmanine E and L and (-)-(R)-leptoclinidamine B (Fig. **9**) displayed similar antidiabetic effects [123].

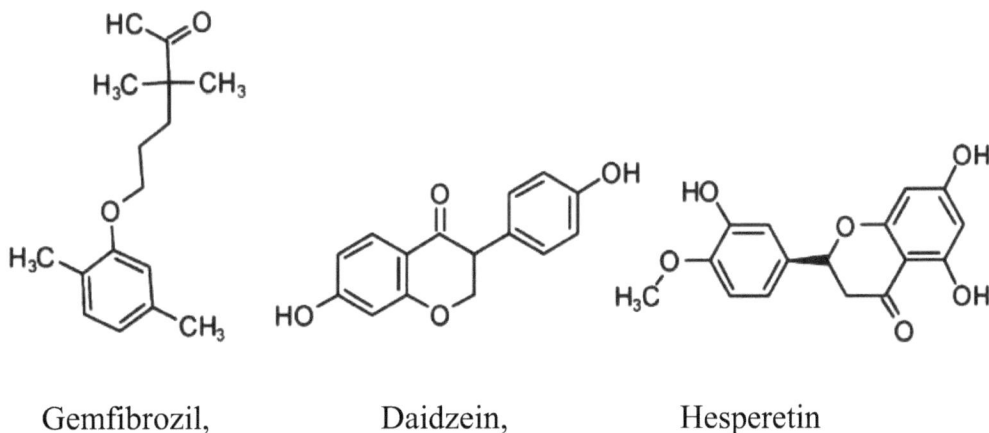

Gemfibrozil,   Daidzein,   Hesperetin

Fig. (5). Gemifibrozil, Daidzein, Hesperetin.

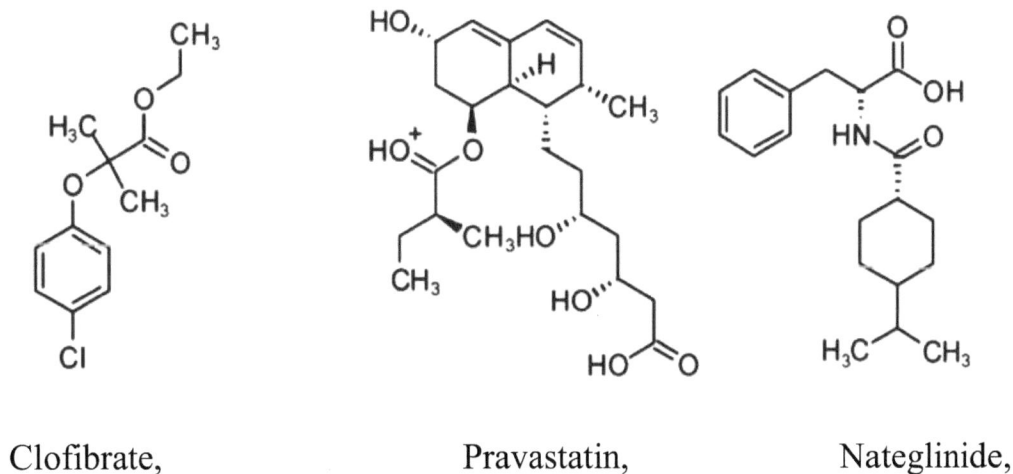

Clofibrate,   Pravastatin,   Nateglinide,

Fig. (6). Clofibrate, Pravastatin, Nateglinide.

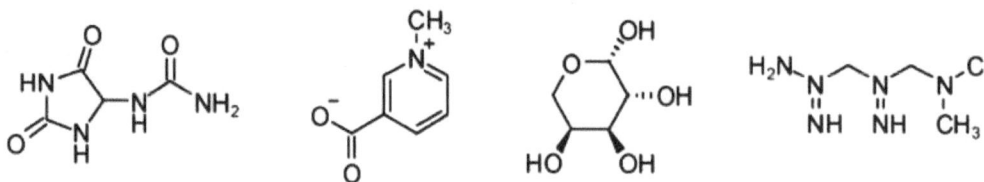

| Allatoin, | Trigonelline, | L-arabinose, | Metformin, |

Fig. (7). Allatoin, Trigonelline, L-arabinose, Metformin.

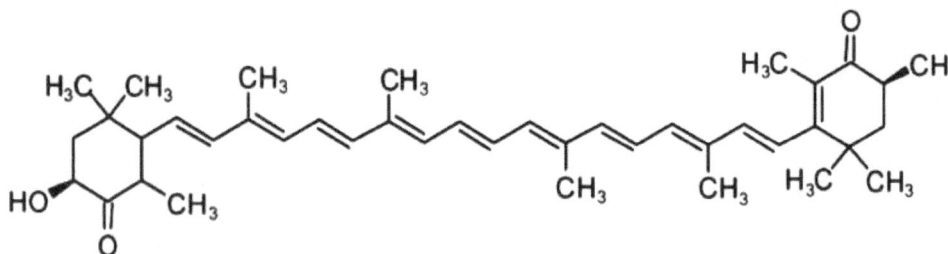

Fig. (8). Astaxanthin.

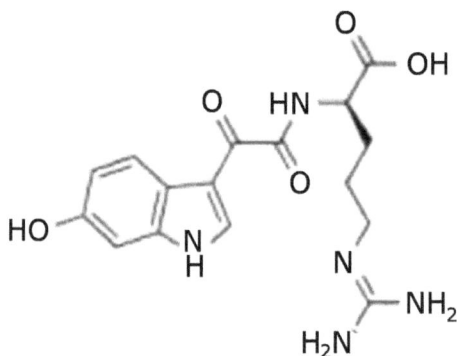

Fig. (9). (-)-(R)-leptoclinidamine B.

Microcosmus exasperatus

Global distribution: This tropical and subtropical species has broad global distribution from all continental waters: Azores, South Africa, Western Mediterranean, California, and Mexico.

Ecology: It is seen on mangrove roots and coral reefs.

Antidiabetics and their mechanisms of action: Alloxan-induced diabetic albino rats administered with the ethanol extracts of this species at a concentration of

were estimated. The results revealed a dose-dependent antidiabetic effect with 200 mg/kg body weight for 14 days showed antidiabetic effects [121] (Fig. **10**).

Fig. (10). *Microcosmus* sp. Sylvain Le Bris; Creative Commons Attribution-Share Alike 4.0 International license. https://commons.wikimedia.org/wiki/File:Microcosmus_sabatieri.jpg.

Microcosmus squamiger

Global distribution: It is native to Australia, and it is also found in the USA, South Africa, Mexico, the Atlantic coast of Spain, and the Mediterranean Sea.

Ecology: It inhabits shallow, littoral rocky habitats.

Antidiabetics and their mechanisms of action: The methanol and ethyl acetate extracts of this species have shown antidiabetic activity by inhibiting the enzyme α-amylase. At the methanol concentrations of 100, 200, and 300 µg/ml, the inhibition percentage values were 40, 60, and 68, respectively , and the corresponding values for the ethyl acetate extracts were 65, 55, and 68, respectively [121] (Fig. **11**).

Microcosmus sp.

Antidiabetics and their mechanisms of action: The extracts of methanol, acetone, and ethyl acetate of this species showed antidiabetic effects by inhibiting the enzyme α-amylase. At the concentrations of 100, 200, and 300 µg/ml, the percentage values of inhibition for the methanol extracts were 48, 50, and 60, respectively ; the corresponding values for the acetone extracts were 29, 50, and, 53, respectively, while for the ethyl acetate extracts, the values were 48, 33, and 67, respectively [121].

Phallusia arabica

Global distribution: Singapore, Indonesia, Philippines, Taiwan, Vietnam, and Madagascar.

Ecology: It is found on artificial substrates such as submerged pontoons of marinas and jetty pilings.

Antidiabetics and their mechanisms of action: The methanol and ethyl acetate extracts of this species have shown antidiabetic activity by inhibiting the enzyme α-amylase. At the methanol concentrations of 100, 200, and 300 µg/ml, the inhibition percentage values were 45, 65, and 68, respectively, and the corresponding values for the ethyl acetate extracts were 45, 60, and 65, respectively [121] (Fig. **11**).

Fig. (11). *Phallusia arabica.* Julien Bidet for MDC Seamarc; Creative CommonsAttribution-Share Alike 4.0 International license. https://commons.wikimedia.org/wiki/File:Ascidies_non_identifi%C3%A9es _Maldives.JPG

Phallusia mammillata

Global distribution: NE Atlantic Ocean, North Sea, Mediterranean Sea, and English Channel.

Ecology: It is found on rocky or muddy substrates in depths up to 200 m.

Antidiabetics and their mechanisms of action: The methanol and ethyl acetate extracts of this species have shown antidiabetic activity by inhibiting the enzyme α-amylase. At the methanol concentrations of 100, 200, and 300 µg/ml, the inhibition percentage values were 65, 78, and 81, respectively, and the corresponding values for the ethyl acetate extracts were 40, 12, and 35, respectively [121] (Fig. **12**).

Fig. **(12).** *Phallusia mammillata.* Diego Delsolicense CC BY-SA; https://commons.wikimedia.org/wiki/File:Pi%C3%B1a_de_mar_%28Phallusia_mammillata%29,_Parque_nat ural_de_la_ Arr%C3%A1bida,_Portugal,_2022-07-19,_DD_50.jpg.

Phallusia nigra

Global distribution: Red Sea, Singapore, and Western Atlantic Ocean.

Ecology: It lives in shallow water and is found attached to hard substrates such as dead corals, pier pilings, or floats (Fig. **13**).

Fig. (13). *Phallusia nigra*. Jorge Stolfi ; Creative Commons Attribution-Share Alike 3.0 Unported license. https://commons.wikimedia.org/wiki/File:Phallusia_nigra_500x500.jpg.

Antidiabetics and their mechanisms of action: An aromatic dicarboxylic acid *viz.* phthalic acid (Fig. **14**) derived from this species displayed a prominent antidiabetic effect [6].

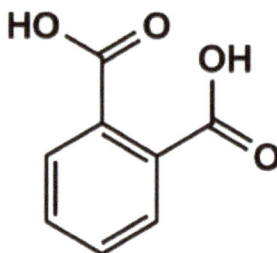

Fig. (14). Phthalic acid.

Polyclinum aurantium

Global distribution: It is native to the NE Atlantic Ocean and ranges from Norway to the Mediterranean Sea.

Ecology: It dwells on rocks and other hard substrates at depths of about 100 m.

Antidiabetics and their mechanisms of action: The methanol and ethyl acetate extracts of this species have shown antidiabetic activity by inhibiting the enzyme α-amylase. At the methanol concentrations of 200 and 300 µg/ml, the inhibition percentage values were 4 and 12, respectively, and the corresponding values for the ethyl acetate extracts were 52 and 65, respectively [121] (Fig. **15**).

Fig. (15). *Polyclinum aurantium.*Matthieu Sontag; Creative Commons Attribution-Share Alike 3.0 Unported, 2.5 Generic, 2.0 Generic and 1.0 Generic license. https://commons.wikimedia.org/wiki/File: Polyclinum_aurantium.jpg.

Polyclinum sp.

Antidiabetics and their mechanisms of action: The ethyl acetate extract of this tunicate species showed antidiabetic effects by inhibiting the enzyme α-amylase. At a concentration of 100 µg/ml, this extract showed inhibition at 44% [121].

Trididemnum savignii

Global distribution: It has a tropical and subtropical distribution globally.

Ecology: This intertidal species is seen at depths of 5-8m.

Antidiabetics and their mechanisms of action : The ethyl acetate extract of this tunicate species showed antidiabetic effects by inhibiting the enzyme α-amylase. At the concentrations of 200 and 300 µg/ml, these extracts showed inhibition percentage values of 30 and 53, respectively [121].

Fish

Elasmobranch Fish

Scyliorhinus canicula

Global distribution: NE Atlantic and the Mediterranean Sea.

Ecology: It occurs both in shallow and deep seas at a depth range of 10- 800m.

Antidiabetics and their mechanisms of action: A biologically active insulin, similar to human insulin, isolated from this shark species has been reported to possess antidiabetic activity by significantly lowering the blood glucose level at a dose of 10 nmol/kg [90] (Fig. **16**).

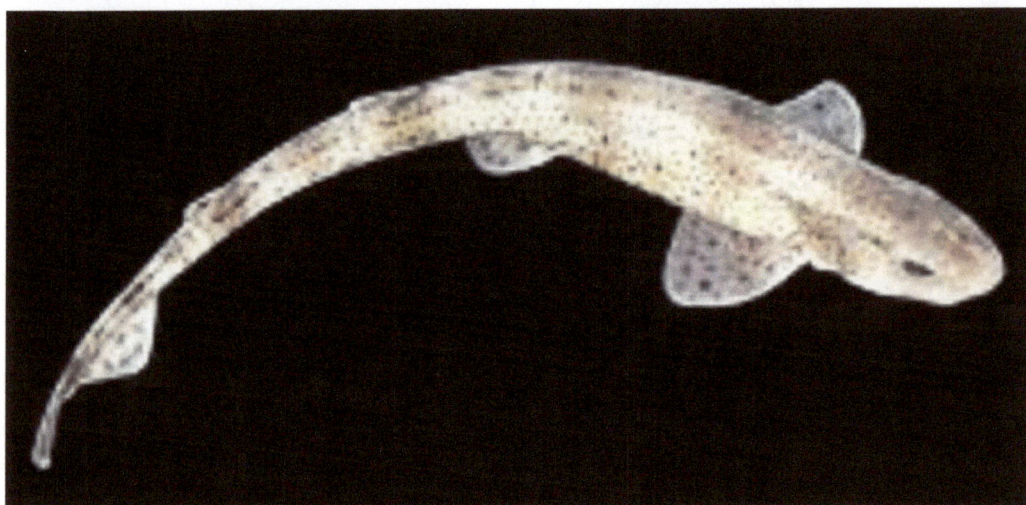

Fig. (16). *Scyliorhinus canicular*. Hans Hillewaert; r the Creative CommonsAttribution-Share Alike 4.0 International license.
https://commons.wikimedia.org/wiki/File:Scyliorhinus_canicula.jpg

Sphyrna lewini

Global distribution: Indo-Pacific: Red Sea, East Africa; Japan to New Caledonia; Eastern Pacific: California, USA to Ecuador.

Ecology: It lives in coastal and offshore waters up to a depth of 1000m.

Antidiabetics and their mechanisms of action: A biologically active insulin, similar to human insulin, isolated from this shark species has been reported to possess antidiabetic activity by significantly lowering the blood glucose level at a dose of 10 nmol/kg [90] (Fig. **17**).

Fig. (17). *Sphyrna lewini.*(Gervais et Boulart; public domainhttps://en.m.wikipedia.org/wiki/File:Sphyrna_ lewini_ Gervais.jpg.

Shark Liver Protein

Antidiabetics and their mechanisms of action: The cholera toxin B subunit and active peptide derived from the shark liver (CTB-APSL) fusion protein have been reported to possess antidiabetic properties. Oral administration of CTB-APSL fusion protein was found to potently reduce the levels of fasting blood glucose and enhance insulin secretion besides improving insulin resistance in type 2 diabetic mice [125].

Teleost Fish

Capros aper

Global distribution: NE Atlantic Ocean and Mediterranean.

Ecology: This demersal species is commonly found on muddy and rocky bottoms or corals.

Antidiabetics and their mechanisms of action: The food protein-derived hydrolysates of this fish have shown *in vitro* dipeptidyl peptidase (DPP-IV) inhibitory activity with an IC50 value of 1.5mg/mL [124] (Fig. **18**).

Oncorhynchus keta

Global distribution: North Pacific: Japan, Korea, Okhotsk, and Bering Sea; Arctic Alaska: California, USA; and Asia: Iran.

Ecology: It inhabits ocean and coastal streams.

Antidiabetics and their mechanisms of action: The administration of oligopeptides derived from this fish in a rat model was found to reduce fasting blood glucose levels (Chellappan *et al.*, 2023). Wong *et al.* (2024) reported that the skin collagen peptides of this species improved glucose metabolism and insulin sensitivity (> 4.5 g/kg/ day) in type 2 DM-affected patients [126] (Fig. **19**).

Fig. (18). *Capros aper*Etrusko25; public domain. https://commons.wikimedia.org/wiki/File:Capros_aper_Sardinia.jpg.

Fig. (19). *Oncorhynchus keta.* Knepp Timothy, U.S. Fish and Wildlife Service; public domain; https://commons.wikimedia.org/wiki/File:Salmon_chum_fish_oncorhynchus_keta.jpg.

Salmo salar

Global distribution: Found in Northern Atlantic and the rivers that flow into it.

Ecology: This anadromous species lives in fresh water for about 3 years of life before migrating to the sea.

Antidiabetics and their mechanisms of action: The hydrolysate-derived peptides from the skin of this fish have shown significant *in vitro* dipeptidyl peptidase (DPP-IV) inhibitory activity (Farias *et al.*, 2022). Further, its food protein-derived hydrolysates have shown *in vitro* dipeptidyl peptidase (DPP-IV) inhibitory activity with an IC50 value of 1.5mg/mL [104] (Fig. **20**).

Fig. (20). *Salmo salar* U.S. National Oceanic and Atmospheric Administration; public domain; https://en.wikipedia.org/wiki/File:Salmo_salar_GLERL_1.jpg.

Sardine pilchardus

Global distribution: NE Atlantic, Mediterranean, and Black Sea.

Ecology: It lives mainly in the upwelling areas of subtropical oceans at depths of 10-100 m.

Antidiabetics and their mechanisms of action: The food protein-derived hydrolysates of this fish have shown *in vitro* dipeptidyl peptidase (DPP-IV) inhibitory activity with an IC50 value of 1.8 mg/mL [104] (Fig. **21**).

Fig. (21). *Sardine pilchardus*Ane2000; Creative CommonsAttribution-Share Alike 4.0 International license. https://commons.wikimedia.org/wiki/File:Sardina_Pilchardus.svg.

Trachinotus ovatus

Global distribution: Mediterranean Sea and Atlantic Ocean (from British Isles and Gulf of Guinea).

Ecology: It is seen in clear waters over mud bottoms or sand. It is known to occasionally enter lagoons and river estuaries. It may also be seen in mid-depths as shoals.

Antidiabetics and their mechanisms of action: The administration of the protein hydrolysates of this fish in streptozotocin-induced diabetic mice displayed a dose-dependent decrease in Fasting Blood Glucose (FBG) by 58%. Further, these hydrolysates have been reported to significantly safeguard the pancreas and kidney in order to ensure the normal secretion of insulin and prevent the occurrence of diabetic nephropathy. Further, the peptides derived from these hydrolysates have also shown DPP-IV inhibitory activity [127] (Fig. **22**).

Fig. (22). *Trachinotus ovatus*. Etrusko25; Creative CommonsAttribution-Share Alike 3.0 Unported license. https://commons.wikimedia.org/wiki/File:Trachinotus_ovatus.jpg.

Zosterisessor ophiocephalus

Global distribution: NE Atlantic, Mediterranean Sea, Azov Sea, and Black Sea.

Ecology: This benthic species is seen in inshore, brackish water, and estuarine areas.

Antidiabetics and their mechanisms of action: The administration of the muscle protein and hydrolysates of this fish to HFHD-induced oxidative stress rats showed significant anti-hyperglycemic effects [6] (Fig. **23**).

Fig. (23). *Zosterisessor ophiocephalus* Jeanne Zaouali; Creative Commons Attribution 3.0 Unported license. https://commons.wikimedia.org/wiki/File:Zosterisessor_ophiocephalus_Tunisia.jpg.

Turtle

Soft-shelled turtle ((*Nilssonia gangetica*), or Ganges softshell turtle) (Fig. **24**).

Fig. (24). Image credit: Cuvier, public domain. https://commons.wikimedia.org/wiki/File: Nilssonia_gangetica.jpg

Global distribution: It is commonly found in South Asia in rivers such as the Ganges, Indus, and Mahanadi.

Ecology: This riverine species lives in both brackish water and freshwater.

Antidiabetics and their mechanisms of action: The food protein-derived hydrolysates of this species exhibited *in vitro* DPP-IV inhibitory activity. Further, the pepsin/trypsin of its yolk has shown similar activity with an IC50 value of 1mg/mL [104].

CONCLUSION

Among the marine chordates, the tunicates have been paid much attention for their natural products with a variety of bioactivities. However, compared to other bioactivities, the antidiabetic effects are known only in limited species, and this warrants intensive antidiabetic investigations on this potent group. Further, while several mechanisms have been reported for the antidiabetic effects of marine fish-derived bioactive compounds, the exact modes of actions and bioavailability of these compounds are yet to be fully understood. Furthermore, most clinical studies are conducted with small sample sizes, which is not sufficient to recommend the usage of these fish-derived bioactive compounds in the treatment and management of T2DM. This calls for more clinical research with a large sample size to further confirm the antidiabetic properties of these bioactive compounds.

CHAPTER 7

Antidiabetics Properties of Marine Fishery By-Products

Abstract: The antidiabetic properties of marine fishery byproducts, such as peptide hydrolysates, collagen peptides, and marine fish oils (ω-3 PUFA) derived from marine fish wastes, are dealt with in this chapter. Further, the possible mechanisms involved in the antidiabetic effects of chitosan and its derivatives from marine crustaceans are also depicted.

Keywords: Antidiabetic effects, Bioactive compounds, Chitosan, Fish oils, Fasting blood glucose, Marine fishery byproducts.

INTRODUCTION

Marine fishery byproducts such as proteins, peptides, and fish oils derived from fish wastes are known to possess antidiabetic, antioxidant, anti-inflammatory, anti-hypertensive, anti-cancer, and immunomodulatory properties. Recent research investigations state that these bioactive compounds could be of great use in the treatment and management of type 2 diabetes mellitus [125].

Bioactive Compounds of Marine Fish and their Antidiabetic Activities

The bioactive proteins, peptides, or liquids derived from the marine fishery byproducts have shown antidiabetic activities through different mechanisms in human and animal models *in vitro* (microorganisms, cell culture, *etc.*). In humans, these compounds reduce fasting blood glucose and enhance the density of beta-cells. In animal models, these compounds are known to inhibit the plasma DPP-IV enzyme activity and enhance plasma insulin level, blood glucose level, *etc.*, and in microorganisms, these bioactive compounds mainly increase insulin release from beta-cells and inhibit DPP-IV activity [125].

Antidiabetic Properties of Marine Fishery by-products

The antidiabetic properties of marine fishery by-products, *viz.* peptide hydrolysates and fish oils, have been reported to show antidiabetic properties in humans [125 - 128].

Antidiabetic Properties of Marine Fish Peptide Hydrolysates

Marine fishery byproduct-derived peptides have been reported to display anti-diabetic activities through several mechanisms, *viz.* stimulating the secretion of Glucagon-Like Peptide 1 (GLP-1), enhancing insulin release, inhibiting DPP-IV activity, increasing glucose uptake, reducing blood glucose levels, and Upregulating Glucose Transporter Type 4 (GLUT4) and peroxisome proliferator-activated receptor alpha (PPAR- α), both of which have been reported to play a crucial role in glucose metabolism and absorption. The administration of marine peptides has shown increased glucose digestion and insulin sensitivity in experimental rats with type 2 diabetes mellitus. These activities are due to the capacity of peptides to reduce the effects of oxidative stress and inflammation, as well as the enhanced expression of GLUT4 and PPAR-α. The peptides derived from the hydrolysate of the muscle of the raw sardine (*Sardine pilchardus*) have been reported to show DPP-IV inhibitory activity with an IC50 value of 1.83 mg/ml. Similarly, insulin and glucagon-like peptide 1 (GLP-1) secreted from the BRIN-BD11 and GLUTag cells of the muscle of the blue whiting fish (*Micromesistius poutassou*) showed inhibition of DPP-IV activity. The peptides from the trimmings of salmon (*Salmo salar*) displayed DPP-IV inhibitory activity with an IC50 value of 0.08 mg/ml. The peptides of the muscle of the boarfish, *Capros aper*, increase insulin secretion, GLP-1 secretion, and glucose tolerance besides inhibiting DPP-IV with an IC50 value of 1.18 mg/ml [128]. The bioactive properties of fish waste-derived peptides are shown in Fig. (**1**).

Antidiabetic Properties of Marine Fish Peptides in Humans with T2DM

The marine fish collagen peptides have been reported to show several antidiabetic effects in humans with T2DM. Such effects include reduced levels of fasting blood glucose, increased levels of insulin sensitivity and secretion index, and an increase in adiponectin [125], as shown in Table **1**.

Antidiabetic properties of marine fish collagen peptides in humans with T2DM [125].

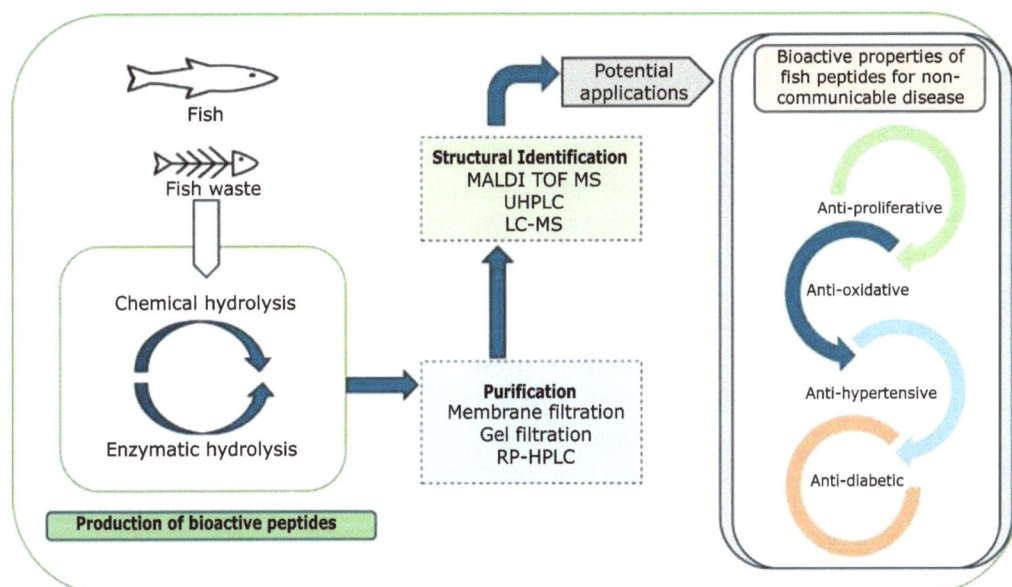

Fig. (1). Bioactive properties of fish waste-derived peptides.

Image credit: Ravi Baraiya, R. Anandan, K. Elavarasan, Patekar Prakash, Sanjaykumar Karsanbhai Rathod, S. R. Radhika Rajasree and V. Renuka (Reproduced with permission)

Table 1. Antidiabetic properties of marine fish collagen peptides in humans with T2DM [125].

Fish Peptide	Human Parameters	Administration (Dose/Duration)	Antidiabetic Effects
Marine fish collagen peptides	T2DM; Healthy	6.5 g twice/day; 3 months	Reduced levels of fasting blood glucose; increased insulin sensitivity index
--do--	T2DM; age: 21-50	2.5 or 5 g once/day; 3 months	Reduction in fasting blood glucose
--do--	T2DM; age: 21-50	10 g/day, 3 months	Reduction in fasting blood glucose
--do---	T2DM with/without hypertension	6.5 g/day; 3 months	Reduction in free fatty acids; increase in adiponectin *
--do--	T2DM and primary hypertension	6.5 g twice/day; 3 months	Reduced levels of fasting blood glucose; increased levels of insulin secretion index and insulin sensitivity index

*Adiponectin, a hormone for insulin sensitivity

Antidiabetic Properties of Halibut (Flatfish) Skin Hydrolysates

The <1.5 kDa ultrafiltration fractions obtained from halibut (flatfish) skin hydrolysate displayed *in vitro* DPP-IV inhibitory activity of 38.2% at a sample concentration of 1 mg solid/mL. The daily administration of TSGH for 30 days was more potent in improving glucose tolerance in streptozotocin-induced diabetic rats. Therefore, the warm-water fish skin gelatin, rich in amino acids, has the potential to be the precursor of DPP-IV inhibitor for the improvement of diabetes. Huang *et al.* (2012) stated the isolation of three peptides with DPP-IV inhibitory activity from the juice hydrolysates of tuna fish. These peptides had the amino acid sequences as Pro-Gly-Val-Gly-Gly-Pro-Leu-Gly-Pro Ile-Gly-Pro-Cys-Tyr-Glu (1412.7 Da), Cys-Ala-Tyr-Gln-Trp-Gln-Arg-Pro-Val-Asp-Arg-Ile-Arg (1690.8 Da) and Pro-Ala-Cys-Gly-Gly-Phe-Try-Ile-Ser-Gly-Arg-Pro-Gly (1304.6 Da). These three peptides exhibited the dose-dependent inhibitory activity of DPP-IV with IC50 values of 116.1, 78.0, and 96.4 µM, respectively. The results further suggest that the tuna cooking juice would be a potential material for the development of new antidiabetic drugs [129].

Anti-diabetic Properties of Marine Fish Oils (ω-3 PUFA) for Humans

It has been reported that marine fish oils (ω-3 PUFA) displayed anti-diabetic activities by enhancing the function of β cells and increasing PPAR-γ activity. The activation of PPAR-γ (Peroxisome proliferator-activated receptors) enhances glucose metabolism and causes insulin sensitization [125]. The antidiabetic effects in humans are shown in Table **2**.

Table 2. Antidiabetic properties of marine fish oils in humans with T2DM [125].

Fish Oil	Human Parameters	Administration (Dose/Duration)	Antidiabetic Effects
Fish oil with DHA	T2DM; Age: 30-70 years,	2400 mg/day; 8 weeks	Enhanced PPAR-γ activity (1)
Omega-3 polyunsaturated fatty acids	Adults with T2DM and fatty liver diseases	2g/day; 12 weeks	Enhanced function of β cells
DHA-enriched fish oil	T2DM; Age: 30-70 years	2400 mg/day; 8 weeks	Increased levels of P16 expression and reduced telomerase activity (2)
Cod-liver oil	pregnant women with GDM; average age, 27 years)	500mg/day; 12-16 weeks	Reduced blood glucose level
Fish oil	Obese patients with T2DM	4.0 g/day; 8 weeks	Improved insulin sensitivity

(Table 2) cont.....

Fish Oil	Human Parameters	Administration (Dose/Duration)	Antidiabetic Effects
Fish oil	Women with GDM; Age: 18–40 years	2000 mg/day; 6 weeks	Improved the gene expression of PPAR-γ in peripheral blood mononuclear cells
DHA with fish oil	T2DM patients	2,400 mg/day; 8 weeks	Reduced body fat; PPARγ gene expression
n-3 PUFAs from fish oil	T2DM patients	2.8 g/day; 24 weeks	Reduced blood glucose and increased serum-resisting insulin
Fish n-3 PUFA	T2DM patients	900mg/3times/day; 10 weeks	Increased adiponectin levels (3)
Fish oil	Women with GDM at 24-28 weeks of the pregnancy	1200 mg/day until delivery.	Only little effects on birth weight
LCHP diet with fish ω-3 fatty acids (4)	Newly diagnosed T2D patients	6 g/day; 12 weeks	Reduced fasting glucose

1. (PPAR-γ or PPARG): Peroxisome proliferator-activated receptor gamma; it is also known as the glitazone reverse insulin resistance receptor.
2. p16: Protein and tumor suppressor.
3. Adiponectin: A hormone that helps with insulin sensitivity and inflammation; low levels of adiponectin are associated with type 2 diabetes.
4. LCHP: Low carbohydrate, high protein diet.

Chitosan and its Derivatives

Chitin is a cellulose-like polysaccharide and is found in the exoskeletons of crustaceans, molluscan organs, fungi, insects, and yeasts. It is a partially degraded product from the N-deacetylation of chitin under alkaline conditions. The Degree of Deacetylation (DD) reaching around 50% demonstrates the generation of chitosan from chitin. However, chitosan produced from chitin with different degrees of deacetylation (40-90%) is available in the market. However, it is generally believed that chitosan with lower Molecular Weight (MW) may have stronger bioactivity than higher MW chitosan. Chitosan and its modified derivatives are bioactive polymers that have a wide range of applications in biomedicine, food processing, non-food chemical industries, and agriculture. Due to their unique bioactivities, such as antioxidant, antimicrobial, and anti-inflammatory, chitosan and its modified derivatives are intensively used in biomedical and pharmaceutical fields, such as drug delivery, wound dressing, artificial skin, and biofilms. Apart from biomedical applications, chitosan and its derivatives have attracted researchers for their health benefits, especially in the development and management of diabetes [130].

Antidiabetic Effects of Chitosan and its Derivatives

The major antidiabetic functions of chitosan and its derivatives, *viz.* the antidiabetic effects of chitosan in animal models, the protection of pancreatic β-Cells [130], and improving insulin secretion, are given below:

i. Antidiabetic effects of chitosan in animal models

Chitosan and its derivatives have been known to reduce hyperglycemia in several diabetic models. Chitosan with a molecular weight of 20 kDa was found to significantly delay the increase of blood glucose levels in low-dose STZ-induced slowly progressive diabetic mice and in genetically obese diabetic KK-Ay mice. Similarly, oral administration of chitosan with an MW of 1.5 kDa reduced fasting blood glucose levels and improved glucose tolerance after 8 weeks. Further, a 42-day administration of chitosan with a MW of < 1 kDa reduced the fasting blood glucose level in obese diabetic db/db mice. N-acetylated 97% DD chitosan was found to reduce blood glucose in high-fat-diet-fed mice. Last but not least, chitosan was also found to possess a hypoglycemic effect in alloxan-induced diabetic mice.

ii. Protection of pancreatic β-Cells and improving insulin secretion by chitosan

Chitosan with a molecular mass of 1200 u was found to facilitate insulin release and glycemic control by accelerating the proliferation of β-cells *in vitro* and improving pancreatic function in STZ-induced diabetic rats. Further, chitosan stopped the death of pancreatic islet cells; protected the pancreas against oxidative damage caused by STZ. Furthermore, chitosan has been reported to possess antidiabetic effects in pancreatic INS-1 β-cells by triggering glucose-stimulated insulin release.

Antidiabetic Effects of Chitosan and its Derivatives from the Marine Crustaceans

Chitosan derived from the chitin of lobster, *Panulirus ornatus* (Subphylum: Crustacea), has been reported to show potent α-amylase and βglucosidase inhibitory activities. At a concentration of 1000μg/ml, this chitosan showed the maximum inhibition percentage values of 64 and 43%. It is also suggested that chitosan may be a potent alternate source for the development of antidiabetic drugs [131]. The antidiabetic effects of chitosan and its derivatives from marine crustaceans are shown in Table **3**.

Table 3. Antidiabetic effects of chitosan and its derivatives from marine crustaceans [130].

Exptl. Subjects	Chitosan Type (MM& DD)	Administration	Conc.	Period	Antidiabetic Activity
Human with prediabetes	MW < 1000 Da; DD ?	Oral	1500 mg/day	12 weeks	control of postprandial glucose
Fructose diet-fed rats	MW = 380 kDa; DD = 89.8%	*Via* diet	5%	21 weeks	Improving glucose and lipid metabolism
3T3-L1 preadipocytes and intestinal Caco-2 cells (*in vitro*)	MW < 1000 Da; DD?	Culture medium	1 and 10 mg/mL	---	Increasing glucose uptake
Diabetic mice	MW 20 kDa; DD?	Drinking water	0.05%, 0.2% or 0.8%	11 weeks	Anti-hyperglycemia and anti-hyperinsulinemia;
Diabetic mice	MW < 1000 Da; DD?	*Via* diet	4%	42 days	Reduction in blood glucose
Streptozotocin-induced diabetic rats	MW 1.5 kDa; DD = 86.5%	Oral	500 and 1000 mg/kg	8 weeks	Anti-hyperglycemia; protecting β-cells; improvement of glucose toleranc

MW: molecular weight; DD: degree of deacetylation

CONCLUSION

Several *in vitro* and *in vivo* studies have displayed the antidiabetic potential of sea fish-derived proteins and peptides. Further, the benefits of sea fish oil that is rich in MUFAs and PUFAs on glycemic control and the improvement of insulin sensitivity are also well known. However, studies on the antidiabetic effects of sea fish oils have been very limited due to the smaller number of marine fish species. Further, intensive research on this aspect would largely help strengthen the potential for sea fish oil to be developed as a dietary supplement for diabetes treatment and management in the future. Furthermore, chitosan and its derivatives from marine crustaceans like shrimps, crabs, and lobsters have been reported to improve glucose homeostasis in diabetic rodents, and these findings are of great use for diabetic intervention. It is also worth mentioning here that very few species of marine crustaceans have been studied for the potential benefits of their chitosan. Above all, intensive research is urgently needed from both animal studies and clinical trials to evaluate the glucose-regulating effects of chitosan and its derivatives.

REFERENCES

[1]　Availabl From: https://www.intechopen.com/chapters/76001

[2]　Casertano M, Vito A, Aiello A, Imperatore C, Menna M. Natural bioactive compounds from marine invertebrates that modulate key targets implicated in the onset of type 2 diabetes mellitus (T2DM) and its complications. Pharmaceutics 2023; 15(9): 2321.
[http://dx.doi.org/10.3390/pharmaceutics15092321] [PMID: 37765290]

[3]　Bhattacharjee R, Mitra A, Dey B, Pal A. Exploration of anti-diabetic potentials amongst marine species: A mini review. Indo Glob J Pharm Sci 2014; 4(2): 65-73.
[http://dx.doi.org/10.35652/IGJPS.2014.109]

[4]　Available from: https://www.healthline.com/health/difference-between-type-1-and-ype-2-diabetes#How-does-diabetes-affect-the-body

[5]　Lankatillake C, Huynh T, Dias DA. Understanding glycaemic control and current approaches for screening antidiabetic natural products from evidence-based medicinal plants. Plant Methods 2019; 15(1): 105.
[http://dx.doi.org/10.1186/s13007-019-0487-8] [PMID: 31516543]

[6]　Chellappan DK, Chellian J, Rahmah NSN, *et al.* Hypoglycaemic molecules for the management of diabetes mellitus from marine sources. Diabetes Metab Syndr Obes 2023; 16: 2187-223.
[http://dx.doi.org/10.2147/DMSO.S390741] [PMID: 37521747]

[7]　Lordan S, Ross RP, Stanton C. Marine bioactives as functional food ingredients: potential to reduce the incidence of chronic diseases. Mar Drugs 2011; 9(6): 1056-100.
[http://dx.doi.org/10.3390/md9061056] [PMID: 21747748]

[8]　Santhanam R, Ramesh S, Shivakumar G. Biology and ecology of pharmaceutical marine cnidarians (series: Biology and ecology of pharmaceutical marine life). (CRC Press (Taylor & Francis). 2019.

[9]　Wang H. Fu Z ming, Han C chao. The potential applications of marine bioactives against diabetes and obesity. Am J Mar Sci 2014; 2(1): 1-8.

[10]　Méresse S, Fodil M, Fleury F, Chénais B. Fucoxanthin, a marine-derived carotenoid from brown seaweeds and microalgae: A promising bioactive compound for cancer therapy. Int J Mol Sci 2020; 21(23): 9273.
[http://dx.doi.org/10.3390/ijms21239273] [PMID: 33291743]

[11]　Zhang H, Tang Y, Zhang Y, *et al.* Fucoxanthin: A promising medicinal and nutritional ingredient. Evid Based Complement Alternat Med 2015; 2015(1): 1-10.
[http://dx.doi.org/10.1155/2015/723515] [PMID: 26106437]

[12]　Ambati R, Phang SM, Ravi S, Aswathanarayana R. Astaxanthin: sources, extraction, stability, biological activities and its commercial applications-a review. Mar Drugs 2014; 12(1): 128-52.
[http://dx.doi.org/10.3390/md12010128] [PMID: 24402174]

[13]　Tang L, Xiao M, Cai S, Mou H, Li D. Potential application of marine fucosyl-polysaccharides in regulating blood glucose and hyperglycemic complications. Foods 2023; 12(13): 2600.
[http://dx.doi.org/10.3390/foods12132600] [PMID: 37444337]

[14]　Albert BB, Derraik JGB, Brennan CM, *et al.* Supplementation with a blend of krill and salmon oil is associated with increased metabolic risk in overweight men. Am J Clin Nutr 2015; 102(1): 49-57.
[http://dx.doi.org/10.3945/ajcn.114.103028] [PMID: 26016867]

[15]　Sebastian K. Nutritional outlook. 2018. Available from: https://www.nutritionaloutlook.com/view/krill-oil-more-effective-reducing-blood-glucose-levels-consuming-fish-alone-says-new-study.

[16]　Bansal MK, Agrawal PK, Gautam A, Singh AK, Nigam AK. Role of krill oil supplementation in

diabetic neuropathy patients – A randomized control trial Med res chronicles 2014; 1(2): 222-33. 2014.

[17]   Lin Q, Guo Y, Li J, He S, Chen Y, Jin H. Antidiabetic effect of collagen peptides from *Harpadon nehereus* bones in streptozotocin-induced diabetes mice by regulating oxidative stress and glucose metabolism. Mar Drugs 2023; 21(10): 518. https://api.semanticscholar.org/CorpusID:263302342
        [http://dx.doi.org/10.3390/md21100518] [PMID: 37888453]

[18]   Santhanam R, Ramesh S, Suleria HAR. Biology and ecology of pharmaceutical marine plants (Series: biology and ecology of pharmaceutical marine life). CRC Press (Taylor & Francis) 2018.

[19]   Senthilkumar R, John SA. Hypoglycaemic activity of marine cyanobacteria in alloxan induced diabetic rats. Pharmacologyonline 2008; 2: 704-14.

[20]   Priatni S, Budiwati TA, Ratnaningrum D, *et al.* Antidiabetic screening of some Indonesian marine cyanobacteria collection. Biodiversitas (Surak) 2016; 17: 2.
        [http://dx.doi.org/10.13057/biodiv/d170236]

[21]   Sæther T, Paulsen SM, Tungen JE, *et al.* Synthesis and biological evaluations of marine oxohexadecenoic acids: PPARα/γ dual agonism and anti-diabetic target gene effects. Eur J Med Chem 2018; 155: 736-53.
        [http://dx.doi.org/10.1016/j.ejmech.2018.06.034] [PMID: 29940464]

[22]   An SM, Cho K, Kim ES, Ki H, Choi G, Kang NS. Description and characterization of the *Odontella aurita* OAOSH22, a marine diatom rich in eicosapentaenoic acid and fucoxanthin, isolated from osan harbor, Korea. Mar Drugs 2023; 21(11): 563.
        [http://dx.doi.org/10.3390/md21110563] [PMID: 37999387]

[23]   Sharifuddin Y, Chin YX, Lim PE, Phang SM. Potential bioactive compounds from seaweed for diabetes management. Mar Drugs 2015; 13(8): 5447-91.
        [http://dx.doi.org/10.3390/md13085447] [PMID: 26308010]

[24]   Ezzat SM, Bishbishy MHE, Habtemariam S, *et al.* Looking at marine-derived bioactive molecules as upcoming anti-diabetic agents: A special emphasis on PTP1B inhibitors. Molecules 2018; 23(12): 3334.
        [http://dx.doi.org/10.3390/molecules23123334] [PMID: 30558294]

[25]   Barde SR, Sakhare RS, Kanthale SB, Chandak PG, Jamkhande PG. Marine bioactive agents: A short review on new marine antidiabetic compounds. Asian Pac J Trop Dis 2015; 5(S1): S209-13.
        [http://dx.doi.org/10.1016/S2222-1808(15)60891-X]

[26]   Kaur M, Bhatia S, Gupta U, *et al.* Microalgal bioactive metabolites as promising implements in nutraceuticals and pharmaceuticals: Inspiring therapy for health benefits. Phytochem Rev 2023; 22(4): 903-33.
        [http://dx.doi.org/10.1007/s11101-022-09848-7] [PMID: 36686403]

[27]   Mayer AMS, Pierce ML, Howe K, *et al.* Marine pharmacology in 2018: Marine compounds with antibacterial, antidiabetic, antifungal, anti-inflammatory, antiprotozoal, antituberculosis and antiviral activities; affecting the immune and nervous systems, and other miscellaneous mechanisms of action. Pharmacol Res 2022; 183(1): 106391.
        [http://dx.doi.org/10.1016/j.phrs.2022.106391] [PMID: 35944805]

[28]   Pandey S, Sree A, Dash SS, Sethi DP, Chowdhury L. Diversity of marine bacteria producing beta-glucosidase inhibitors. Microb Cell Fact 2013; 12(1): 35.
        [http://dx.doi.org/10.1186/1475-2859-12-35] [PMID: 23590573]

[29]   Artanti N, Maryani F, Mulyani H, Dewi R, Saraswati V, Murniasih T. Bioactivities screening of indonesian marine bacteria isolated from sponges.In: Annales Bogorienses. 2016; pp. 23-8.

[30]   Amr K, Ibrahim N, Elissawy AM, Singab ANB. Unearthing the fungal endophyte *Aspergillus terreus* for chemodiversity and medicinal prospects: a comprehensive review. Fungal Biol Biotechnol 2023; 10(1): 6.

[http://dx.doi.org/10.1186/s40694-023-00153-2] [PMID: 36966331]

[31] Aboutabl ME, Maklad YA, Abdel-Aziz MS, Abd El-Hady FK. *In vitro* and *in vivo* studies of the antidiabetic potential of Red Sea sponge-associated fungus "*Aspergillus unguis*" isolate SP51-EGY with correlations to its chemical composition. J Appl Pharm Sci 2022; 12(8): 165-78.
[http://dx.doi.org/10.7324/JAPS.2022.120817]

[32] Seo C, HanYim J, Kum Lee H, Oh H. PTP1B inhibitory secondary metabolites from the Antarctic lichen *Lecidella carpathica*. Mycology 2011; 2(1): 18-23.
[http://dx.doi.org/10.1080/21501203.2011.554906]

[33] Gunathilaka T, Rangee Keertihirathna L, Peiris D. Advanced pharmacological uses of marine algae as an anti-diabetic therapy. Nat Med Plants 2022; 11: 79.
[http://dx.doi.org/10.5772/intechopen.96807]

[34] Mandlik RV, Naik SR, Zine S, Ved H, Doshi G. Antidiabetic activity of *Caulerpa racemosa*: Role of proinflammatory mediators, oxidative stress, and other biomarkers. Planta Med Int Open 2022; 9(1): e60-71.
[http://dx.doi.org/10.1055/a-1712-8178]

[35] Unnikrishnan PS, Jayasri MA. Antidiabetic studies of *Chaetomorpha antennina* extract using experimental models. J Appl Phycol 2017; 29(2): 1047-56.
[http://dx.doi.org/10.1007/s10811-016-0991-4]

[36] Qin L, Yang Y, Hao J, *et al.* Antidiabetic-activity sulfated polysaccharide from *Chaetomorpha linum*: Characteristics of its structure and effects on oxidative stress and mitochondrial function. Int J Biol Macromol 2022; 207: 333-45.
[http://dx.doi.org/10.1016/j.ijbiomac.2022.02.129] [PMID: 35227705]

[37] Lee SH, Jeon YJ. Anti-diabetic effects of brown algae derived phlorotannins, marine polyphenols through diverse mechanisms. Fitoterapia 2013; 86: 129-36.
[http://dx.doi.org/10.1016/j.fitote.2013.02.013] [PMID: 23466874]

[38] Pham T, Nguyen T, Yun H, *et al.* Echinochrome a prevents diabetic nephropathy by inhibiting the pkc-iota pathway and enhancing renal mitochondrial function in db/db Mice. Mar Drugs 2023; 21(4): 222.
[http://dx.doi.org/10.3390/md21040222] [PMID: 37103361]

[39] Abirami RG, Kowsalya S. Antidiabetic activity of *Ulva fasciata* and its impact on carbohydrate metabol- ism enzymes in alloxan induced diabetic rats. Int J Res Phytochem Pharmacol 2013; 3: 136-41.

[40] Nwosu F, Morris J, Lund VA, Stewart D, Ross HA, McDougall GJ. Anti-proliferative and potential anti-diabetic effects of phenolic-rich extracts from edible marine algae. Food Chem 2011; 126(3): 1006-12. https://www.sciencedirect.com/science/article/pii/S0308814610015293
[http://dx.doi.org/10.1016/j.foodchem.2010.11.111]

[41] Gunathilaka TL, Samarakoon K, Ranasinghe P, Peiris LDC. Antidiabetic potential of marine brown algae—a mini review. J Diabetes Res 2020; 2020: 1-13.
[http://dx.doi.org/10.1155/2020/1230218] [PMID: 32377517]

[42] Gunathilaka T, Rangee Keertihirathna L, Peiris D. Advanced pharmacological uses of marine algae as an anti-diabetic therapy. Nat Med Plants 2022; 11:79.

[43] Sharma A, Koneri R, Jha DK. A review of pharmacological activity of marine algae in Indian coast. IJPSR 2019; 10(8): 3540-9.

[44] Rushdi MI, Abdel-Rahman IAM, Attia EZ, *et al.* The biodiversity of the genus *Dictyota*: Phytochemical and pharmacological natural products prospectives. Molecules 2022; 27(3): 672.
[http://dx.doi.org/10.3390/molecules27030672] [PMID: 35163940]

[45] Lee SH, Kang SM, Ko SC, *et al.* Octaphlorethol A: A potent α-glucosidase inhibitor isolated from Ishige foliacea shows an anti-hyperglycemic effect in mice with streptozotocin-induced diabetes. Food

Funct 2014; 5(10): 2602-8.
[http://dx.doi.org/10.1039/C4FO00420E] [PMID: 25145393]

[46] Moheimanian N, Mirkhani H, Purkhosrow A, Sohrabipour J, Jassbi AR. *In Vitro* and *In Vivo* antidiabetic, α-glucosidase inhibition and antibacterial activities of three brown algae, *Polycladia myrica, Padina antillarum*, and *Sargassum boveanum*, and a Red Alga, *Palisada perforata* from the Persian Gulf. Iran J Pharm Res 2023; 22(1): e133731.
[http://dx.doi.org/10.5812/ijpr-133731] [PMID: 38116547]

[47] Alvarado-Sansininea JJ, Tavera-Hernández R, Jiménez-Estrada M, *et al*. Antibacterial, antidiabetic, and toxicity effects of two brown algae:*Sargassum buxifolium* and *Padina gymnospora*. Int J Plant Biol 2022; 14(1): 63-76.
[http://dx.doi.org/10.3390/ijpb14010006]

[48] Nagi MA, Mohamed SA. Antidiabetic and antioxidative potential of *Cystoseira myrica*. Biochemistry 2014; 8(6): 197-208.

[49] Oliyaei N, Moosavi-Nasab M, Tamaddon AM, Tanideh N. Antidiabetic effect of fucoxanthin extracted from *Sargassum angustifolium* on streptozotocin-nicotinamide-induced type 2 diabetic mice. Food Sci Nutr 2021; 9(7): 3521-9.
[http://dx.doi.org/10.1002/fsn3.2301] [PMID: 34262712]

[50] Available from: https://www.google.com/search?q=antidiabetic+properties+of+Sargassum+fusiforme.

[51] Jia RB, Li ZR, Wu J, *et al*. Antidiabetic effects and underlying mechanisms of anti-digestive dietary polysaccharides from *Sargassum fusiforme* in rats. Food Funct 2020; 11(8): 7023-36.
[http://dx.doi.org/10.1039/D0FO01166E] [PMID: 32716443]

[52] Payghami N, Jamili S, Rustaiyan A, Saeidnia S, Nikan M, Gohari AR. Alpha-amylase inhibitory activity and sterol composition of the marine algae, *Sargassum glaucescens*. Pharmacognosy Res 2014; 7(4): 314-21.
[PMID: 26692744]

[53] Lee YH, Kim HR, Yeo MH, *et al*. Anti-diabetic potential of *Sargassum horneri* and *Ulva australis* extracts *In Vitro* and *In Vivo*. Curr Issues Mol Biol 2023; 45(9): 7492-512.
[http://dx.doi.org/10.3390/cimb45090473] [PMID: 37754257]

[54] Gotama TL, Husni A, Ustadi . Antidiabetic activity of *Sargassum hystrix* extracts in streptozotocin-induced diabetic rats. Prev Nutr Food Sci 2018; 23(3): 189-95.
[http://dx.doi.org/10.3746/pnf.2018.23.3.189] [PMID: 30386746]

[55] Xie X, Chen C, Fu X. Modulation effects of *Sargassum pallidum* extract on hyperglycemia and hyperlipidemia in type 2 diabetic mice. Foods 2023; 12(24): 4409.
[http://dx.doi.org/10.3390/foods12244409] [PMID: 38137213]

[56] Kawamura-Konishi Y, Watanabe N, Saito M, *et al*. Isolation of a new phlorotannin, a potent inhibitor of carbohydrate-hydrolyzing enzymes, from the brown alga *Sargassum patens*. J Agric Food Chem 2012; 60(22): 5565-70.
[http://dx.doi.org/10.1021/jf300165j] [PMID: 22594840]

[57] Unnikrishnan PS, Suthindhiran K, Jayasri MA. Antidiabetic potential of marine algae by inhibiting key metabolic enzymes. Front Life Sci 2015; 8(2): 148-59.
[http://dx.doi.org/10.1080/21553769.2015.1005244]

[58] Ohta T, Sasaki S, Oohori T, Yoshikawa S, Kurihara H. Alpha-glucosidase inhibitory activity of a 70% methanol extract from ezoishige (*Pelvetia babingtonii* de Toni) and its effect on the elevation of blood glucose level in rats. Biosci Biotechnol Biochem 2002; 66(7): 1552-4.
[http://dx.doi.org/10.1271/bbb.66.1552] [PMID: 12224640]

[59] Lee YS, Shin KH, Kim BK, Lee S. Anti-Diabetic activities of fucosterol from *Pelvetia siliquosa*. Arch Pharm Res 2004; 27(11): 1120-2.
[http://dx.doi.org/10.1007/BF02975115] [PMID: 15595413]

[60] Vuppalapati L, Velayudam R, Nazeer Ahamed KFH, Cherukuri S, Kesavan BR. The protective effect of dietary flavonoid fraction from *Acanthophora spicifera* on streptozotocin induced oxidative stress in diabetic rats. Food Sci Hum Wellness 2016; 5(2): 57-64.
[http://dx.doi.org/10.1016/j.fshw.2016.02.002]

[61] Sabarianandh JV, Subha V, Manimekalai K. Antidiabetic activity of red marine algae *in vitro*: A review. Annals of SBV 2020; 9(1): 22-6.
[http://dx.doi.org/10.5005/jp-journals-10085-8117]

[62] Ramesh S, Santhanam R, Sankar V. Marine biopharmaceuticals: Scope and prospects bentham books. 2024; p. 350.

[63] Sanger G, Rarung LK, Damongilala LJ, Kaseger BE, Montolalu L A DY. Phytochemical constituents and antidiabetic activity of edible marine red seaweed (*Halymenia durvilae*). IOP Conf Ser Earth Environ Sci 2019; 278(1): 012069.
[http://dx.doi.org/10.1088/1755-1315/278/1/012069]

[64] Zhang Y, Wu S. Hypoglycemic effect of polysaccharides from *Porphyra yezoensis* associated with reduced intestinal α-amylase activity in diabetes mellitus KKAy Mice. J Aquat Food Prod Technol 2022; 31(10): 1109-14.
[http://dx.doi.org/10.1080/10498850.2022.2133583]

[65] Ben Abdallah Kolsi R, Ben Salah H, Jardak N, *et al.* Effects of *Cymodocea nodosa* extract on metabolic disorders and oxidative stress in alloxan-diabetic rats. Biomed Pharmacother 2017; 89: 257-67.
[http://dx.doi.org/10.1016/j.biopha.2017.02.032] [PMID: 28235688]

[66] Gono CMP, Ahmadi P, Hertiani T, Septiana E, Putra MY, Chianese G. A comprehensive update on the bioactive compounds from seagrasses. Mar Drugs 2022; 20(7): 406.
[http://dx.doi.org/10.3390/md20070406] [PMID: 35877699]

[67] Popov AM, Krivoshapko ON. Protective effects of polar lipids and redox-active compounds from marine organisms at modeling of hyperlipidemia and diabetes. J Biomed Sci Eng 2013; 6(5): 543-50.
[http://dx.doi.org/10.4236/jbise.2013.65069]

[68] Das SK, Samantaray D, Sahoo SK, Patra JK, Samanta L, Thatoi H. Bioactivity guided isolation and structural characterization of the antidiabetic and antioxidant compound from bark extract of *Avicennia officinalis* L. S Afr J Bot 2019; 125: 109-15.
[http://dx.doi.org/10.1016/j.sajb.2019.07.011]

[69] Sachithanandam V, Lalitha P, Parthiban A, Mageswaran T, Manmadhan K, Sridhar R. A review on antidiabetic properties of indian mangrove plants with reference to island ecosystem. Evid Based Complement Alternat Med 2019; 2019(1): 1-21.
[http://dx.doi.org/10.1155/2019/4305148] [PMID: 31885647]

[70] Widiastuti EL, Ardiansyah BK, Nurcahyani N, Silvinia A. Antidiabetic potency of jeruju (*Acanthus ilicifolius* L.) ethanol extract and taurine on histopathological response of mice kidney (Mus musculus L.) induced by alloxan. Journal of Physics: Conference Series 2021; 12052.
[http://dx.doi.org/10.1088/1742-6596/1751/1/012052]

[71] Derebe D, Wubetu M, Alamirew A. Hypoglycemic and antihyperglycemic activities of 80% methanol root extract of *Acanthus polystachyus* Delile (Acanthaceae) in type 2 diabetic rats. Clin Pharmacol 2020; 12: 149-57.
[http://dx.doi.org/10.2147/CPAA.S273501] [PMID: 33061672]

[72] Ramanathan T, Gurudeeban S, Satyavani K, Balasubramanian T. Antidiabetic effect of a black mangrove species *Aegiceras corniculatum* in alloxan-induced diabetic rats. J Adv Pharm Technol Res 2012; 3(1): 52-6.
[http://dx.doi.org/10.4103/2231-4040.93560] [PMID: 22470894]

[73] Al-Jaghthmi OHA, Zeid IEDMEAA, Al-Ghamdi KMS, Heba HM, Ahmad MS. Antihyperglycemic,

antioxidant and antiapoptotic effect of *Rhizophora mucronata* and *Avicennia marina* in streptozotocin-induced diabetic rats. Med Arch (Sarajevo, Bosnia Herzegovina) 2020; 74(6): 421-7.

[74]   Gayen S, Jana S, Das Gupta B, *et al.* Exploration of anti-diabetic activity and metabolite profiling of *Bruguiera cylindrica* (l.) Bl.— *in vivo* anti-diabetic activity, exploration of molecular mechanism, and network pharmacological analysis. J Pharm Pharmacol 2024; 76(7): 798-812.
[http://dx.doi.org/10.1093/jpp/rgae030] [PMID: 38546509]

[75]   Karimulla SK, Kumar BP. Anti diabetic and Anti hyperlipidemic activity of bark of *Bruguiera gymnorrhiza* on streptozotocin induced diabetic rats. Asian J Pharm Sci Tech 2011; 1(1): 4-7.

[76]   Bui TT, Nguyen KPT, Nguyen PPK, Le DT, Nguyen TLT. Anti-inflammatory and α-glucosidase inhibitory activities of chemical constituents from *Bruguiera parviflora* leaves. J Chem 2022; 2022(1): 1-9.
[http://dx.doi.org/10.1155/2022/3049994]

[77]   Nabeel MA, Kathiresan K, Manivannan S. Antidiabetic activity of the mangrove species *Ceriops decandra* in alloxan-induced diabetic rats. J Diabetes 2010; 2(2): 97-103.
[http://dx.doi.org/10.1111/j.1753-0407.2010.00068.x] [PMID: 20923491]

[78]   Biswas B, Golder M, Devnath HS, Mazumder K, Sadhu SK. Comparative antihyperglycemic, analgesic and anti-inflammatory potential of ethanolic aerial root extracts of *Ceriops decandra* and *Ceriops tagal*: Supported by molecular docking and ADMET analysis. Heliyon 2023; 9(3): e14254.
[http://dx.doi.org/10.1016/j.heliyon.2023.e14254] [PMID: 36938384]

[79]   Thirumurugan G. +5. Evaluation of antidiabetic activity of *Excoecaria agallocha* L. in Alloxan Induced Diabetic Mice Nar Prod 2010; Na72560934.

[80]   Ansari P, Flatt PR, Harriott P, Abdel-Wahab YHA. Insulin secretory and antidiabetic actions of *Heritiera fomes* bark together with isolation of active phytomolecules. PLoS One 2022; 17(3): e0264632.
[http://dx.doi.org/10.1371/journal.pone.0264632] [PMID: 35239729]

[81]   Lakshmi V, Kumar R, Srivastava AK. Antidiabetic effect of leaves of *Kandelia candel* Linn. Nat Prod 2013; 9(8): 319-21.

[82]   Yusoff NA, Yam MF, Beh HK, *et al.* Antidiabetic and antioxidant activities of *Nypa fruticans* Wurmb. vinegar sample from Malaysia. Asian Pac J Trop Med 2015; 8(8): 595-605.
[http://dx.doi.org/10.1016/j.apjtm.2015.07.015] [PMID: 26321511]

[83]   Andrade-Cetto A, Escandón-Rivera SM, Torres-Valle GM, Quijano L. Phytochemical composition and chronic hypoglycemic effect of *Rhizophora mangle* cortex on STZ-NA-induced diabetic rats. Rev Bras Farmacogn 2017; 27(6): 744-50.
[http://dx.doi.org/10.1016/j.bjp.2017.09.007]

[84]   Usman , Amir MM, Erika F, *et al.* Antidiabetic activity of leaf extract from three types of mangrove originating from sambera coastal region Indonesia. Research Journal of Pharmacy and Technology 2019; 12(4): 1707-12.
[http://dx.doi.org/10.5958/0974-360X.2019.00284.1]

[85]   Adhikari A, Ray M, Das A, Sur T. Antidiabetic and antioxidant activity of *Rhizophora mucronata* leaves (Indian sundarban mangrove): An *in vitro* and *in vivo* study. Ayu 2016; 37(1): 76-81.
[http://dx.doi.org/10.4103/ayu.AYU_182_15] [PMID: 28827960]

[86]   Camara AK, Baldé ES, Haidara M, *et al. In vivo* antidiabetic activities of aqueous extract of *Anchomanes difformis* (Blume) Eng, *Rhizophora racemosa* G. Mey and *Ravenala madagascariensis* sonn. European J Med Plants 2020; 31(18): 15-22.
[http://dx.doi.org/10.9734/ejmp/2020/v31i1830339]

[87]   Hasan MN, Sultana N, Akhter MS, Billah MM, Islamp KK. Hypoglycemic effect of methanolic extract of *Sonneratia caseolaris* (fruits) – A mangrove plant from Bagerhat region, the. J Innov Dev Strateg 2013; 7(1): 1-6.

[88] Elya B, Budiarso FS, Hanafi M, Gani MA, Prasetyaningrum PW. Two tetrahydroxyterpenoids and a flavonoid from *Xylocarpus moluccensis* M.Roem. and their α-glucosidase inhibitory and antioxidant capacity. J Pharm Pharmacogn Res 2024; 12(3): 453-76.
[http://dx.doi.org/10.56499/jppres23.1816_12.3.453]

[89] Anjum K, Abbas SQ, Shah SAA, Akhter N, Batool S, Hassan SS. Marine sponges as a drug treasure. Biomol Ther (Seoul) 2016; 24(4): 347-62.
[http://dx.doi.org/10.4062/biomolther.2016.067] [PMID: 27350338]

[90] Santhanam R, Ramesh S, Sunilson AJ. Biology and ecology of pharmaceutical marine sponges (Series: biology and ecology of pharmaceutical marine life). CRC Press (Taylor & Francis) 2018.

[91] Ajabnoor MAM, Tilmisany AK, Taha AM, Antonius A. Effect of Red Sea sponge extracts on blood glucose levels in normal mice. J Ethnopharmacol 1991; 33(1-2): 103-6.
[http://dx.doi.org/10.1016/0378-8741(91)90169-E] [PMID: 1943158]

[92] Francis P, Chakraborty K. Clathriketal, a new tricyclic spiroketal compound from marine sponge *Clathria prolifera* attenuates serine exopeptidase dipeptidyl peptidase-IV. Nat Prod Res 2022; 36(12): 3069-77.
[http://dx.doi.org/10.1080/14786419.2021.1956491] [PMID: 34315292]

[93] Yamazaki H, Nakazawa T, Sumilat DA, Takahashi O, Ukai K, Takahashi S. Euryspongins A-C, three new unique sesquiterpenes from a marine sponge *Euryspongia* sp. Bioorganic Med Chem Lett 2013; 23(7): 2151–4. https://www.sciencedirect.com/science/article/pii/S0960894X13001352

[94] Tiwari P, Rahuja N, Kumar R, Lakshmi V, Srivastava MN, Agarwal SC. Search for antihyperglycemic activity in few marine flora and fauna. Indian J Sci Technol 2008; 1(5): 1-5.
[http://dx.doi.org/10.17485/ijst/2008/v1i5.4]

[95] Abdjul DB, Yamazaki H, Takahashi O, Kirikoshi R, Ukai K, Namikoshi M. Isopetrosynol, a new protein tyrosine phosphatase 1B inhibitor, from the marine sponge *Halichondria* cf. *panicea* collected at Iriomote Island. Chem Pharm Bull (Tokyo) 2016; 64(7): 733-6.
[http://dx.doi.org/10.1248/cpb.c16-00061] [PMID: 27373628]

[96] Ansarizadeh A, Kafilzadeh F, Tamadoni Jahromi S, Kargar M, Gozari M. Isolation, identification and evaluation of the anti-diabetic activity of secondary metabolites extracted from bacteria associated with the Persian Gulf sponges (*Haliclona* sp. and *Niphates* sp.). Iran J Fish Sci 2023; 22(3): 511-25.

[97] Liu Y, Palaniveloo K, Alias SA, Sathiya Seelan JS. Species diversity and secondary metabolites of sarcophyton-associated marine fungi. Molecules 2021; 26(11): 3227.
[http://dx.doi.org/10.3390/molecules26113227] [PMID: 34072177]

[98] Gao Y, Zhang X, Yuan J, Zhang C, Li S, Li F. CRISPR/Cas9-mediated mutation on an insulin-like peptide encoding gene affects the growth of the ridgetail white prawn *Exopalaemon carinicauda*. Front Endocrinol (Lausanne) 2022; 13: 986491.
[http://dx.doi.org/10.3389/fendo.2022.986491] [PMID: 36246877]

[99] El-Desoky MS, Takeuchi R, Katayama H, Tsutsui N. Chemical synthesis of insulin-like peptide 1 and its potential role in vitellogenesis of the kuruma prawn *Marsupenaeus japonicus*. J Pept Sci 2023; 29(12): e3529.
[http://dx.doi.org/10.1002/psc.3529] [PMID: 37403818]

[100] Jiang Q, Zheng H, Zheng L, *et al.* Molecular characterization of the insulin-like androgenic gland hormone in the swimming crab, *Portunus trituberculatus*, and its involvement in the insulin signaling system. Front Endocrinol (Lausanne) 2020; 11: 585.
[http://dx.doi.org/10.3389/fendo.2020.00585] [PMID: 32982976]

[101] Farias TC, de Souza TSP, Fai AEC, Koblitz MGB. Critical Review for the Production of Antidiabetic Peptides by a Bibliometric Approach. Nutrients 2022; 14(20): 4275.
[http://dx.doi.org/10.3390/nu14204275] [PMID: 36296965]

[102] Santhanam R, Gopinath M, Ramesh S. Biology and ecology of pharmaceutical marine molluscs

(series: Biology and ecology of pharmaceutical marine life). CRC Press (Taylor & Francis). 2019.

[103] Joy M, Chakraborty K, Pananghat V. Comparative bioactive properties of bivalve clams against different disease molecular targets. J Food Biochem 2016; 40(4): 593-602.
[http://dx.doi.org/10.1111/jfbc.12256]

[104] Nong NTP, Hsu JL. Characteristics of food protein-derived antidiabetic bioactive peptides: A literature update. Int J Mol Sci 2021; 22(17): 9508.
[http://dx.doi.org/10.3390/ijms22179508] [PMID: 34502417]

[105] Salas S, Chakraborty K. Functional properties of the marine gastropod molluscs *Chicoreus ramosus* and *Babylonia spirata* collected from the southern coast of india. J Aquat Food Prod Technol 2020; 29(3): 264-78.
[http://dx.doi.org/10.1080/10498850.2020.1722776]

[106] Jenivi A , Jenivi A, Thilaga RD. Evaluation of antidiabetic activity of the marine gastropod, *Turbinella pyrum* from Gulf of Mannar, India. J Pharm Negat Results 2022; 1006-9.
[http://dx.doi.org/10.47750/pnr.2022.13.S06.134]

[107] Chakraborty K, Joy M. Anti-diabetic and anti-inflammatory activities of commonly available cephalopods. Int J Food Prop 2017; 20(7): 1655-65.
[http://dx.doi.org/10.1080/10942912.2016.1217008]

[108] Krishnan S, Chakraborty K, Joy M. First report of chromenyl derivatives from spineless marine cuttlefish *Sepiella inermis* : Prospective antihyperglycemic agents attenuate serine protease dipeptidyl peptidase-IV. J Food Biochem 2019; 43(5): e12824.
[http://dx.doi.org/10.1111/jfbc.12824] [PMID: 31353519]

[109] Abd El Hafez MSM, Aziz Okbah MAE, Ibrahim HAH, Hussein AAER, El Moneim NAA, Ata A. First report of steroid derivatives isolated from starfish *Acanthaster planci* with anti-bacterial, anti-cancer and anti-diabetic activities. Nat Prod Res 2022; 36(21): 5545-52.
[http://dx.doi.org/10.1080/14786419.2021.2021200] [PMID: 34969331]

[110] Santhanam R, Ramesh S, David SRN. Biology and ecology of pharmaceutical marine life: Echinoderms (Series: Biology and ecology of pharmaceutical marine life). CRC Press (Taylor & Francis) 2019.

[111] Marmouzi I, Tamsouri N, El Hamdani M, Attar A, Kharbach M, Alami R. Pharmacological and chemical properties of some marine echinoderms. Rev Bras Farmacogn 2018; 28(5): 575-81.
[http://dx.doi.org/10.1016/j.bjp.2018.05.015]

[112] Soleimani S, Pirmoradloo E, Farmani F, Moein S, Yousefzadi M. Antidiabetic and antioxidant properties of sea urchin *Echinometra mathaei* from the persian gulf. J Kerman Univ Med Sci 2021; 28(1): 104-15.

[113] Moreno-García DM, Salas-Rojas M, Fernández-Martínez E, *et al.* Sea urchins: an update on their pharmacological properties. PeerJ 2022; 10: e13606.
[http://dx.doi.org/10.7717/peerj.13606] [PMID: 35811815]

[114] Bharathi VR, Prabha YS. Screening of *in-vitro* Antidiabetic Potentials from the Protein Extracts of *Stomopneustes variolaris.* Uttar Pradesh J Zool 2023; 44(18): 18-29.
[http://dx.doi.org/10.56557/upjoz/2023/v44i183600]

[115] Gelani C, La Villa C, Duhaylungsod GL, Indanao RJ, Labrador Q, Mamalo M. Anti-inflammatory and anti-diabetic potential of echinoderm species from mindanao, philippines. GA – 70th Annu Meet. 88(15): 195.

[116] Raj S. M.H A, Gothandam KM, Rameshan AT. Chemical composition, *in-vitro* Antioxidant, antibacterial, and anti-algal activity of lakshadweep feather star spp. *Comaster Schlegelii* and *Himerometra Robustipinna.* Thalassas 2022; 38(1): 183-96.
[http://dx.doi.org/10.1007/s41208-021-00366-5]

[117] Chen Y, Wang Y, Yang S, Yu M, Jiang T, Lv Z. Glycosaminoglycan from *Apostichopus japonicus*

improves glucose metabolism in the liver of insulin resistant mice. Mar Drugs 2019; 18(1): 1.
[http://dx.doi.org/10.3390/md18010001] [PMID: 31861309]

[118]   Xia X, Qi J, Liu Y, *et al.* Bioactive isopimarane diterpenes from the fungus, *Epicoccum* sp. HS-1, associated with *Apostichopus japonicus.* Mar Drugs 2015; 13(3): 1124-32.
[http://dx.doi.org/10.3390/md13031124] [PMID: 25738327]

[119]   Mamalo MI, Gelani CD, Lavilla CA Jr. The polar fraction of a Sea cucumber (*Bohadschia argus*) is a potential source of anti-inflammatory and hypoglycemic agents. Int J Biosci 2022; 20(6): 254-60.

[120]   Wang T, Zheng L, Wang S, Zhao M, Liu X. Anti-diabetic and anti-hyperlipidemic effects of sea cucumber (*Cucumaria frondosa*) gonad hydrolysates in type II diabetic rats. Food Sci Hum Wellness 2022; 11(6): 1614-22.
[http://dx.doi.org/10.1016/j.fshw.2022.06.020]

[121]   Prabhu AS, Ananthan G. Alpha-amylase inhibitory activities of ascidians in the treatment of diabetes mellitus. Bangladesh J Pharmacol 2014; 9(4): 498-500.
[http://dx.doi.org/10.3329/bjp.v9i4.20017]

[122]   Zhu Y, Gao H, Han S, Li J, Wen Q, Dong B. Antidiabetic activity and metabolite profiles of ascidian *Halocynthia roretzi.* J Funct Foods 2022; 93: 105095.
[http://dx.doi.org/10.1016/j.jff.2022.105095]

[123]   Santhanam R, Ramesh S. Biology and ecology of pharmaceutical marine tunicates (Series: Biology and ecology of pharmaceutical marine life). CRC Press (Taylor & Francis). 2019.

[124]   Meenakshi VK, Gomathy S, Senthamarai S, Paripooranaselvi M, Chamundeswari KP, College APCM. Hepatoprotective activity of the ethanol extract of simple ascidian, *Microcosmus exasperatus* Heller, 1878. European J Zool Res 2013; 2(4): 32-8.

[125]   Zhou X, Chai L, Wu Q, Wang Y, Li S, Chen J. Anti-diabetic properties of bioactive components from fish and milk. J Funct Foods 2021; 85: 104669.
[http://dx.doi.org/10.1016/j.jff.2021.104669]

[126]   Wong RPM, Zhou ZK, Strappe PM. The anti-obesogenic and anti-diabetic properties of marine collagen peptides. Front Food Sci Technol 2024; 3: 1270392.
[http://dx.doi.org/10.3389/frfst.2023.1270392]

[127]   Wan P, Cai B, Chen H, *et al.* Antidiabetic effects of protein hydrolysates from *Trachinotus ovatus* and identification and screening of peptides with α-amylase and DPP-IV inhibitory activities. Curr Res Food Sci 2023; 6: 100446. https://www.sciencedirect.com/science/article/pii/S266592712300014X
[http://dx.doi.org/10.1016/j.crfs.2023.100446] [PMID: 36816000]

[128]   Baraiya R, Anandan R, Elavarasan K, *et al.* Potential of fish bioactive peptides for the prevention of global pandemic non-communicable disease: production, purification, identification, and health benefits. Discover Food 2024; 4(1): 34.
[http://dx.doi.org/10.1007/s44187-024-00097-5]

[129]   Wang TY, Hsieh CH, Hung CC, Jao CL, Chen MC, Hsu KC. Fish skin gelatin hydrolysates as dipeptidyl peptidase IV inhibitors and glucagon-like peptide-1 stimulators improve glycaemic control in diabetic rats: A comparison between warm- and cold-water fish. J Funct Foods 2015; 19: 330-40.
https://www.sciencedirect.com/science/article/pii/S1756464615004545
[http://dx.doi.org/10.1016/j.jff.2015.09.037]

[130]   Tzeng HP, Liu SH, Chiang MT. Antidiabetic properties of chitosan and its derivatives. Mar Drugs 2022; 20(12): 784.
[http://dx.doi.org/10.3390/md20120784] [PMID: 36547931]

[131]   Arasukumar B, Prabakaran G. B G. Evaluation of *In Vitro* antidiabetic effect of chitosan from lobster (*Panulirus ornatus*) shell. Int J Pharm Biol Sci 2019; 19(3): 12-5.

SUBJECT INDEX

Spasmolytic activities 45
Sterols, oxidized 228
Stress, endoplasmic reticulum 1

T

Therapeutic 14, 72
   agent 72
   effects 14
Treatment, systemic glucocorticoid 244
Tumor suppressor 262

V

Vanadium-containing proteins (VCPs) 234

W

Waters coastal 25, 48, 51, 57, 118, 126, 139,
      160, 190, 207, 216, 224
   marine 207
   muddy 216
Western Atlantic Ocean 75, 146, 248